Bioethics and Social Ethics in The Modern World

"Bioethics is an interdisciplinary science bridging healthcare, philosophy, social sciences, economics, and environmental studies. The book "BIOETHICS AND SOCIAL ETHICS IN THE MODERN WORLD: THE ENVIRONMENTAL AND SOCIAL SUSTAINABILITY CONTEXT(S)" highlights in a most erudite manner from an unbiased standpoint, giving readers a most intelligible and informative overview of the subject."

—Prof. Michael Cheng-tek Tai, Ph.D., *Chungshan Medical University, Taiwan University of Saskatchewan, Canada (Adjunct professor)*

"Ethics is at the very heart of humanity. The book "BIOETHICS AND SOCIAL ETHICS IN THE MODERN WORLD: THE ENVIRONMENTAL AND SOCIAL SUSTAINABILITY CONTEXT(S)" is an invaluable resource for anyone seeking to understand the ethical dimensions of sustainability in health-care, economics, and environmental studies. It is a compelling anthology that tackles a diverse range of ethical challenges in modern society. The chapters span foundational philosophical discussions as well as the topics at the intersections of law, media influence, and sustainability. I can recommend this book as both a critical resource and an engaging read for those interested in bioethics, social justice, and the sustainability of human and environmental systems."

—Prof. Ana Marušić, M.D., Ph.D., *Department of Research in Biomedicine and Health, Center for Evidence-based Medicine, University of Split, School of Medicine, Split, Croatia*

"The book brings together different aspects of responsible and sustainable practices in healthcare, economics, and environmental studies. It intertwines theoretical discussions on fundamental bioethical concepts and principles with their application in contemporary society and market economy, trying to find different novel perspectives on the general philosophical and ethical foundations based on the idea of human dignity. Exploring new concepts and challenges, it never loses sight of the basic fundamentals of ethical theory."

—Prof. Ana Borovečki, M.D., Ph.D., *Professor of Philosophy and Comparative Literature, Specialist in Clinical Pharmacology with Toxicology, School of Public Health "Andrija Štampar" Faculty of Medicine, University of Zagreb, Croatia*

"This book offers an important exploration of the intersection between ethics, sustainability, and contemporary societal challenges. Its relevance to today's world lies in addressing pressing issues such as environmental degradation, social justice, and sustainable development from a philosophical and ethical standpoint. The strength of the book lies in its combination of ethical theory with practical applications, such as corporate social responsibility, sustainable healthcare, and responsible marketing.

The book emphasizes the need for collaboration between policymakers, businesses, and communities and proposes actionable strategies for sustainable development that respect human dignity. It is a valuable resource for academics, policymakers, and practitioners navigating the ethical challenges of a rapidly evolving world."

—Prof. Sunčica Oberman Peterka, Ph.D., *Professor of Entrepreneurship and Strategic management, Faculty of Economics and Business, Josip Juraj Strossmayer University of Osijek, Croatia*

Ivan Pavić · Nikša Alfirević · Suzana Vuletić
Editors

Bioethics and Social Ethics in The Modern World

The Environmental and Social Sustainability Context(s)

palgrave
macmillan

Editors
Ivan Pavić
University of Split
Split, Croatia

Suzana Vuletić
Catholic Faculty of Theology
University of Osijek
Đakovo, Croatia

Nikša Alfirević
Faculty of Economics, Business & Tourism
University of Split
Split, Croatia

ISBN 978-3-031-86420-9 ISBN 978-3-031-86418-6 (eBook)
https://doi.org/10.1007/978-3-031-86418-6

This Palgrave Macmillan imprint is published by the registered company Springer Nature Switzerland AG
The registered company address is: Gewerbestrasse 11, 6330 Cham, Switzerland

We dedicate this volume to the memory of Professor Emeritus Luka Tomasevic of the Faculty of Catholic Theology, University of Split, whose devotion to clinical bioethics helped establish the International Society for Clinical Bioethics on firm intellectual and moral foundations. It is to his vision of a more humane and sustainable world that we owe much of the spirit in which this book was conceived, and it is in his honor that we humbly offer these pages.

Preface

In today's rapidly changing world, new social, environmental, and technological challenges to our traditional ethical assumptions are continuously emerging. Starting with an assumption that the principles of bioethics and social ethics could serve as a good starting point for transdisciplinary solutions to those challenges, we edited this volume, *Bioethics, and Social Ethics in The Modern World: The Environmental and Social Sustainability Context(s)*, as a source of reference, motivation, and inspirations for decision-makers.

Although contributors to this volume belong to diverse fields of theology, economics, law, and sustainability, they share a commitment to address the significant ethical challenges of our time. Together, they examine environmental ethics, healthcare ethics, different aspects of sustainable development, and other relevant topics.

Many of the chapters grew from the 20th International Scientific Conference of the International Society for Clinical Bioethics, and we have also included additional scholarly work to broaden the discussion. By blending theoretical insights with practical considerations, this volume looks at different ways bioethics and social ethics might assist us in reaching environmental and social sustainability.

Split, Croatia Ivan Pavić
Split, Croatia Nikša Alfirević
Đakovo, Croatia Suzana Vuletić

Acknowledgments Publishing this volume has been made possible with the assistance of the University of Split, Croatia, and the town of Makarska, Croatia.

Contents

1	**Introduction** Ivan Pavić, Nikša Alfirević, and Suzana Vuletić	1
2	**Toward an Integrative Framework of Bioethics, Social Ethics, and Sustainability** Nikša Alfirević, Ivan Pavić, and Tea Franić	5
3	**An Era of Mass Production of Evil** Tsuyoshi Awaya	15
4	**Bioethical Principles in the Vortex of Law** Pavo Barišić	23
5	**Aristotle, Aquinas, and Integral Bioethics** Filip Veber	37
6	**Edmund Pellegrino's Proposition of Patient-Centered Virtue Ethics** Szczepan Kaleciak	47
7	**Ethical and Legal Dilemmas of Surrogacy: *Gain for One, Loss for None?*** Nevena Aljinović and Ana Jeličić	59
8	**Future of the Abortion in Croatia** Ivana Tucak, Josip Berdica, and Laura Seleš	71

9 **Gender Refugees and Sex Reassignment Surgery: Medical Tourism from Japan to Thailand** 97
Keisuke Shishido

10 **Using Traditional, Complementary, and Alternative Medicine Amidst the COVID-19 Pandemic: Results of a Database Search Focusing on Diverse Approaches and Usage** 105
Yutaka Kato

11 **Green Consumer Ethics and Values—Which Role Does the Media Play in Presenting and Shaping Individual Opinions of Ethics and Sustainability?** 115
Isabell Koinig

12 **Ethical Determinants of Consumer Behavior and Responsible Marketing Practice** 133
Danijela Križman and Darko Rendulić

13 **Food Consumption Patterns and Sustainability: Results of Empirical Research** 147
Mirna Leko Šimić, Marija Čutura, and Marina Stanić

14 **Dalmatian Cuisine: A Traditional Facilitator of Ethics and Sustainability** 167
Vedran Poljak, Mateja Dugandžić, Eva Pavić, and Valentina Rahelić

15 **Catholic Ecology and the Contribution of the Franciscan Tradition to Integral Ecology** 177
Luka Tomašević and Suzana Vuletić

16 **Environmental Sustainability in Times of Economic Crises: Ethical and Macroeconomic Dimensions** 187
Lena Malešević Perović and Maja Mihaljević Kosor

17 **Religious Tourism and the Environment: World Youth Day 2023 and Pope Francis' Environmental Catechesis** 199
Edward R. Grant and Marija Barić Đurđević

18 Medical and Social Risks of Health Tourism 209
Silvana Pavlinović, Suzana Vuletić, and Gordana Pelčić

19 Conclusion 219
Ivan Pavić, Nikša Alfirević, and Suzana Vuletić

Index 221

Notes on Contrbutors

Nikša Alfirević is a Professor at the Faculty of Economics, Business and Tourism, University of Split, Croatia. His research focuses on education management, leadership, and sustainable development education, with publications addressing management practices and educational reforms.

Nevena Aljinović is an Assistant Professor at the University Department of Forensic Sciences, University of Split, Croatia. Her research focuses on criminal law, procedural law, organized crime, and criminology with victimology.

Tsuyoshi Awaya is a Professor of Law at Okayama Shoka University, Japan, specializing in medical law and bioethics. His research focuses on bioethics, self-determination, medical research, and biotechnology.

Pavo Barišić is a Professor of Philosophy at the Faculty of Croatian Studies, University of Zagreb. His research interests include the history of Croatian philosophy and fundamental problems of philosophy.

Josip Berdica is a Professor at the Faculty of Law, University of Osijek, Croatia, specializing in social and political philosophy. His research focuses on the political philosophy of John Rawls and the intersection of society and law.

Marija Čutura is a Professor at the Faculty of Economics, University of Mostar, Bosnia and Herzegovina. Her research interests include marketing strategies, consumer behavior, and brand management.

Mateja Dugandžić is affiliated with the University Study Program in Hotel Management and Gastronomy in Makarska, University of Split, Croatia. Her research focuses on hospitality management, gastronomy, and tourism development.

Marija Barić Đurđević, MA is an adjunct assistant at the University of Split, Croatia. She graduated from the University of Zagreb, double majoring in Latin language, Croatian Latinity, Roman literature, and Philosophy. She is currently a Ph.D. student in Philosophy at the Faculty of Humanities and Social Sciences, University of Zagreb.

Tea Franić is a seasoned international sustainable development professional with over two decades in the United Nations, CGIAR, and the World Bank focusing on sustainable finance, partnerships, project and policy advisory, and external relations.

Edward R. Grant, J.D., M.A., has widely published and lectured extensively in the field of law and medical ethics since the 1980s. He currently serves as an assistant professor at the Pellegrino Center for Clinical Bioethics at Georgetown University and as a clinical ethics consultant at Medstar Georgetown U. Hospital in Washington DC.

Ana Jeličić is an Assistant Professor at the University Department of Forensic Sciences, University of Split, Croatia. Her research includes bioethical and criminal law responses to human organ trafficking, as well as forensic profiling and behavioral evidence analysis.

Szczepan Kaleciak is affiliated with the Pontifical University of John Paul II in Krakow, Poland. His research interests include theology, philosophy, and bioethics.

Yutaka Kato is a Professor at Shiga University of Medical Science. His research focuses on the roles and challenges of autonomy and conscience in the medical field within a culturally diverse world.

Isabell Koinig is affiliated with the University of Klagenfurt. Her research interests include communication studies, with a focus on advertising and consumer behavior.

Maja Mihaljević Kosor is an Associate Professor at the Faculty of Economics, Business and Tourism, University of Split, Croatia. Her research interests encompass the economics of education, higher education efficiency analysis, public sector economics, and sustainable development goals.

Danijela Križman is a Professor at the Faculty of Economics and Tourism "Dr. Mijo Mirković" at Juraj Dobrila University of Pula, Croatia. Her research focuses on tourism destination marketing, service marketing management, and sustainable marketing.

Eva Pavić is affiliated with the Clinical Hospital Center Zagreb and the University of Applied Health Sciences, Zagreb, Croatia. Her research interests include clinical nutrition, dietetics, and public health.

Ivan Pavić is a Professor Emeritus at the Faculty of Economics, Business and Tourism, University of Split. His research focuses on microeconomics, market concentration, and industrial organization.

Silvana Pavlinović is associated with the Study of Hotel Management and Gastronomy in Makarska, University of Split, Croatia. Her work contributes to the development of hospitality and gastronomy education to enhance the quality of tourism services in the region.

Gordana Pelčić is an Assistant Professor at the Faculty of Medicine, University of Rijeka, Croatia. Her research interests include pediatrics, medical ethics, bioethics, clinical bioethics, and religion, as well as decision-making in pediatric care.

Lena Malešević Perović is a Professor at the Faculty of Economics, Business and Tourism, University of Split, Croatia. Her research focuses on macroeconomics, economic growth, and the impact of fiscal policies on economic development.

Vedran Poljak is an Associate Professor at the University Department of Marine Studies, University of Split. His research interests include nutrition research, marine ecology, fisheries management, and marine conservation.

Valentina Rahelić is associated with the University Hospital Center Zagreb and the University of Applied Health Sciences, Zagreb, Croatia. Her research focuses on endocrinology, diabetes management, and metabolic disorders.

Darko Rendulić is affiliated with the University of Applied Sciences Karlovac, Croatia. His research interests include socially and environmentally responsible marketing and supply chain practices, sustainable development, and supply chain management.

Laura Seleš is a Ph.D. candidate at the Faculty of Law, of the Josip Juraj Strossmayer University in Osijek, Croatia. Her research focuses on the role of morality, justice, and fairness in law, particularly how these principles form the foundation of legal values.

Keisuke Shishido is a Professor of Law at Okayama University, Japan. His research focuses on medical law and bioethics, with a particular emphasis on organ transplants and assisted reproductive technology.

Mirna Leko Šimić is a Professor at the Faculty of Economics and Business Osijek, Josip Juraj Strossmayer University of Osijek. Her research focuses on marketing, consumer behavior, and market research.

Marina Stanić is an Associate Professor at the Faculty of Economics and Business in Osijek, Josip Juraj Strossmayer University of Osijek, Croatia. Her research interests include human behavior and decision-making processes in various business and personal situations, especially in the management context.

Luka Tomašević is a Professor Emeritus with the Catholic Faculty of Theology, University of Split, Croatia. His research includes moral theology, bioethics, environmental ethics, and the implications of modern technology, framed through the theoretical lens of Franciscan theology.

Ivana Tucak is a Professor at the Faculty of Law, University of Osijek, Croatia, specializing in legal theory. Her research covers theories of rights, constitutional rights, bioethics, human rights, and public health law.

Filip Veber is a Slovenian Catholic theologian and visiting scholar at the Pontifical University of Santa Croce (Rome). Concurrently, he is also completing his advanced studies at the Pontifical University of St. Thomas Aquinas (Angelicum). His research focuses on the teachings of St. Thomas Aquinas, with an emphasis on Thomistic Christology (the glorification of Christ) and theological anthropology (nature and grace).

Suzana Vuletić is a Professor of Moral Theology and Clinical Bioethics at the Catholic Faculty of Theology in Đakovo, Josip Juraj Strossmayer University of Osijek, Croatia. Her research interests include moral theology, bioethics, and medical ethics, particularly focusing on clinical dilemmas in healthcare and the ethical implications of modern technology for human dignity, the sanctity of life, and its inviolability.

List of Figures

Fig. 13.1 Sustainable food consumption patterns (Source: Authors) 157
Fig. 14.1 Gastronomy as a reason for visiting Dalmatia 171
Fig. 14.2 Presentation of traditional dishes that guests tasted during their stay in Dalmatia 172
Fig. 14.3 Overall rating of Dalmatian cuisine 172

List of Tables

Table 8.1	Number of legally induced abortions in Croatia (1990–2023)	76
Table 8.2	Three Views of Abortion	79
Table 13.1	Sample characteristics (N = 547)	154
Table 13.2	Dimensions of sustainable food knowledge construct	155
Table 13.3	Dimensions of the sustainable food consumption construct	156
Table 13.4	Behavioral patterns of sustainable food consumption	156
Table 13.5	Respondents' knowledge about sustainable food (total)	158
Table 13.6	Respondents' knowledge about sustainable food (by questions)	159
Table 13.7	Correlation analysis	159
Table 16.1	The definitions and sources of the variables	193
Table 16.2	Descriptive statistics of variables	193
Table 16.3	Results of the model	194

CHAPTER 1

Introduction

Ivan Pavić, Nikša Alfirević, and Suzana Vuletić

Abstract In an age of relentless biotechnological intervention and consumerist excess, urgent ethical and environmental questions demand our attention. Bioethics, informed by social, legal, and political thought, provides the critical perspective needed to confront the fallout of unchecked scientific ambition and resource depletion. Drawing on contributions from the 20th International Scientific Conference of the International Society for Clinical Bioethics, this volume provides insights from theology, economics, law and environmental studies to inspire more just and sustainable approaches to human dignity and planetary stewardship.

I. Pavić
University of Split, Split, Croatia
e-mail: ivan.pavic@efst.hr

N. Alfirević (✉)
Faculty of Economics, Business and Tourism, University of Split, Split, Croatia
e-mail: nalf@efst.hr

S. Vuletić
Catholic Faculty of Theology, Josip Juraj Strossmayer University of Osijek, Đakovo, Croatia

I. Pavić et al. (eds.), *Bioethics and Social Ethics in The Modern World*,
https://doi.org/10.1007/978-3-031-86418-6_1

Keywords Biotechnology · Environment · Bioethics · Consumerism · Sustainability

The twenty-first century is characterized by tendencies of human prosperity and scientific curiosity of biotechnological interventions, which subordinated the natural reality to the despotic mastery of contemporary man without regard to ethical consequences. Subjective pragmatic interests, hedonism, consumerism, and productivism have become the focus of contemporary society, which remains indifferent to social injustice and environmental damage.

Bioethics has become increasingly relevant as it encompasses social ethics, legal and political theories and addresses the application of ethics to life's challenges (Casado & Castán, 2018). Bioethics could, therefore, become an intellectual framework of choice for researchers interested in the trans-disciplinary analysis of human life and the ethical implications of various human activities (Cockburn & Cundill, 2018) and technological innovations (Hostiuc et al., 2016).

Consequently, in contemporary bioethical discourse, the world is seen as dangerously compromised by social and environmental challenges. As a result of such irresponsible conduct, characterized by not respecting the intrinsic value of life, limited natural resources, and the inability to renew, an existential crisis emerges. The expansion of science and technology has not only caused profound changes in nature but has also become an appeal to change human consciousness and basic attitudes toward the world as a part of our responsibility to preserve the social and natural heritage for future generations.

Starting with the perspective that the issues of social and environmental sustainability cannot be considered only from a technical-scientific or economic perspective, we edited this volume, which aims to evaluate the issues of bioethics and sustainability from cultural, ethical, and religious perspectives. We also aim to bridge the gap between the philosophical analysis of attitudes and values and sustainable behaviors in healthcare, social sciences, economics, and environmental studies.

In the volume, we have selected the papers previously discussed at the *20th International Scientific Conference of the International Society for Clinical Bioethics*, entitled "*Integral Bioethics, Health Tourism and Protection of the Environment*." Along with conference participants from Japan,

Thailand, the United States, Poland, Austria, Slovenia, and Croatia, we invited additional authors to contribute and structured the volume in four sections:

I .*Bioethics and Social Ethics in the Modern World: Theory and Applications,*
II .*Integral Approach to Bioethics and Its Applications,*
III .*Bioethics and Social Ethics in the Social World: Toward the Social Sustainability,* and
IV .*Bioethics and Social Ethics in the Natural World: Toward the Environmental Sustainability.*

We hope that this volume contributes to synthesizing the topics of bioethics and sustainability into a comprehensive framework that addresses a broad range of sustainability issues and that readers will find it a valuable addition to their libraries of reference bioethics, ethics, and sustainability books.

References

Casado, M., & Castán, M. L. M. (2018). Bioethics and legal philosophy: An inevitable and necessary reflection. In M. J. Falcón y Tella, J. A. Martínez Muñoz (Eds.), *Challenges to Legal Theory - Essays in Honour of Professor José Iturmendi Morales*. Brill.https://doi.org/10.1163/9789004439450_015

Cockburn, J., & Cundill, G. (2018). Ethics in transdisciplinary research: Reflections on the implications of 'science with society'. In: C. Macleod, J. Marx, P. Mnyaka, G. Treharne (eds), *The palgrave handbook of ethics in critical research*. Palgrave Macmillan. https://doi.org/10.1007/978-3-319-74721-7_6

Hostiuc, S., Moldoveanu, A., Dascălu, M. I., et al. (2016). Translational research—the need of a new bioethics approach. *Journal of Translational Medicine, 14*, 16. https://doi.org/10.1186/s12967-016-0773-4

CHAPTER 2

Toward an Integrative Framework of Bioethics, Social Ethics, and Sustainability

Nikša Alfirević, Ivan Pavić, and Tea Franić

Abstract This chapter introduces an integrative model, including insights from bioethics, social ethics, and sustainability concepts, and addresses the multiple ethical challenges in contemporary societies. We explore the moral obligations humans hold toward other living beings and the environment and address the question of environmental ethics. We

Disclaimer Tea Franić contributes to this chapter in their personal capacity, as a sustainable development expert. The opinions expressed in this chapter are the author's own and do not reflect the view of UNDP or the United Nations.

N. Alfirević (✉)
Faculty of Economics, Business and Tourism, University of Split, Split, Croatia
e-mail: nalf@efst.hr

I. Pavić
University of Split, Split, Croatia
e-mail: ivan.pavic@efst.hr

T. Franić
United Nations, New York, USA

I. Pavić et al. (eds.), *Bioethics and Social Ethics in The Modern World*,
https://doi.org/10.1007/978-3-031-86418-6_2

emphasize cooperation among policymakers, national and multinational stakeholders, and the private sector to advance sustainable development toward achieving human dignity and ecological balance.

Keywords Bioethics · Social ethics · Sustainability · Integrative model

Introduction

Contemporary challenges, including climate change, loss of biodiversity, unequal distribution of resources, and other factors degrading the natural and socioeconomic environments, require an integrated ethical approach. While traditional bioethics is concerned with medical issues and human health, it now includes broader ecological and social factors, which concern sustainability's social and environmental dimensions (Beauchamp & Childress, 2013; Jonas, 1984; Potter, 1971, 1988). Ethics examines justice, equity, and shared responsibility. When applied to environmental issues, it emphasizes how environmental burdens and benefits often mirror existing divides by race, class, and region (Bauman, 2011; Rawls, 1971). Moreover, sustainability research focuses on achieving the long-term balance of economic development and environmental protection but also addresses social harmony and resilience (Kates, Parris, & Leiserowitz, 2005; Purvis et al., 2019). These three areas must be aligned and coordinated to address the complexity of contemporary political, economic, and environmental crises.

As ordinary citizens feel threatened by the challenges of peace and democracy, as well as issues of maintaining their standard of living and ensuring a healthy environment, a critical view of mainstream ideas of environmentalism becomes relevant once again. Specifically, the environmental justice movement, described by Sandler and Pezzullo (2007), has criticized environmentalists for favoring the protection of wilderness areas and "exotic" species while paying too little attention to pollution and degradation of the shared environment, in which they spend most of their everyday lives. In addition, other sustainability problems overburden marginalized communities.

Therefore, an integrative model should combine bioethics, social ethics, and sustainability principles. This chapter presents the fundamental ideas underpinning such a model and refers to the capabilities approach to justice and well-being (Nussbaum, 2006; Sen, 1999), intersectional views of vulnerability (Gaard, 2011; Plumwood, 2002), environmental virtue ethics (Sandler & Cafaro, 2005), and the political economy of global environmental governance (Peet & Watts, 2004; Newell, 2019).

It can be helpful to imagine this approach as a web of connected ethical dimensions composed of bioethical principles, social justice goals, and sustainability issues. Those dimensions are in the mutual relationship, and as they value the positions of different stakeholders, a range of ethical standpoints are recognized. This combined outlook is compatible with Norton's concept (2005) of normative methodology, guiding policy, research, and practice without resorting to restricted theoretical positions.

Theoretical Foundations

Evolution of Sustainability Thinking

Building on its origins in resource management, sustainability has expanded to address environmental, social, and economic dimensions (Kates et al., 2005; Purvis et al., 2019). Fundamental sustainability declarations,including the 1972 Stockholm Declaration and the 1987 Brundtland Report show that environmental protection, human well-being,and economic stability are inter-connected and mutually dependent. Over the years, these discussions have taken shape in the model of three sustainability "pillars" (environment, society, economy), introduced at the Johannesburg Summit and applied in the 2015 UN Sustainable Development Goals (SDGs).

However, it can be argued that dominant sustainability frameworks place too much emphasis on economic growth, but rather neglect its ecological and social justice implications (Rasmussen, 1996). The conventional paradigm of perpetual growth has come into question (Meadows et al., 2004), as it does not address well the challenges of natural resource constraints and long-term environmental damage.

In contrast to focus on economic output, the capabilities approach (Sen, 1999; Nussbaum, 2006) centers on individuals'potential to achieve meaningful goals, prioritizing well-being, freedom, and opportunity over financial indicators. This framework shows that effective policy must place value on well-being and human development, instead of looking only to minimum living standards (Norton, 2005).

Bioethics and Sustainability

Bioethics has traditionally been associated with beneficence, nonmaleficence, autonomy, and justice. However, these principles can be extended beyond the healthcare context by shaping policies that influence human populations and the broader natural and social environment (Beauchamp & Childress, 2013).

All sustainability science starts from an ethical imperative not to harm the environment so that future generations may continue to live without unwanted consequences (Jonas, 1984). This imperative is compatible with the main postulates of bioethics. Specifically, beneficence and nonmaleficence call for addressing direct and indirect harm to communities and ecosystems caused by the degradation of the natural environment. Traditionally framed as an individual right, autonomy also has a collective dimension, which becomes evident in resource management and public decision-making. The fair resource allocation debate should focus on justice, as a leading principle in the resolution of climate change concerns and identification of relevant stakeholders.

Jonas's (1984) "precautionary ethic" warns against irreversible damage, including climate instability and biodiversity loss. Applying bioethical reasoning to emerging technologies introduces elements of caution, accountability, and respect for localized autonomy (Grunwald, 2016; Jasanoff, 2004). Consequently, technical knowledge and moral considerations must inform policy choices and technological innovations.

Social Ethics and Collective Responsibility

There is a strong relationship between social ethics and sustainability, as both emphasize shared responsibility and the collective duty to achieve the sustainable development goals, which requires sharing of sustainability costs and fostering inclusive governance (Holifield et al., 2018). Therefore, there should be a focus on marginalized groups and communities and their burden in achieving sustainable development. Building on Rawls's (1971) principles of justice, the environmental justice literature looks at the disproportionate impact of pollution and resource extraction on those already marginalized in the socioeconomic context (Sandler & Pezzullo, 2007). Research on intersectionality, i.e., the socioeconomic characteristics and social stratification of individual opportunities and experiences (Gaard, 2011; Plumwood, 2002), further illustrates how race, social class, and gender play fundamental roles in determining environmental vulnerability and resilience.

Groups with a historical experience of oppression are often associated with excessive experience of environmental harm and injustice originating from oppressive social structures (Birkeland, 1993; Shiva, 1989). According to Bauman (2011), reshaping decision-making systems to incorporate marginalized groups is both necessary from an ethical point of view and practically viable. Recognizing such linkages guides policy strategies toward challenging deeply rooted power imbalances and striving for more equitable social arrangements.

Toward an Integrative Framework

Rationale and Conceptual Basis

Van Rensselaer Potter (1971, 1988) discussed bioethics as a "bridge" linking scientific knowledge and human values, thus serving as a predecessor to the field of environmental ethics. This discussion has been followed by Jonas (1984), who focused on accountability to future generations and the need to exercise ecological caution. In a later development, Jenkins (2013) integrated philosophy, theology, and social thought and raised the point that religious insights and moral reasoning can deepen environmental stewardship.

There is, also, a need to address the systemic inequities, as the Western line of thinking often fails to overlook environmental justice (Sandler & Pezzullo, 2007), by focusing on remote wilderness areas, while ignoring nearby communities facing environmental risks. Such inequities could be addressed by non-Western theoretical points, including the value of indigenous knowledge and community ecological practices (Shiva, 1989) and traditional knowledge systems (Berkes, 1999). Gottlieb (2006) terms this inclusive perspective "cosmopolitan depth," accentuating the rich diversity of moral communities and their contributions to environmental ethics.

Key Dimensions of Integration

Sustainability relies on preserving biodiversity, ecological functions, and limits of economic growth (Meadows et al., 2004). Applying bioethical principles, i.e., beneficence, nonmaleficence, justice, and autonomy

(Beauchamp & Childress, 2013), further refines the existing approach(es) of environmental ethics, focusing on disadvantaged groups and communities. This has been further clarified by Sandler and Pezzullo (2007), who warned that protecting all communities burdened with significant pollution is ultimately an ethical question.

Social ethics draws attention to the issues of fairness, inclusion, and solidarity (Bauman, 2011; Rawls, 1971). Whereas mainstream environmental discourse centers on ecosystems, environmental justice spotlights the spaces people inhabit daily—communities,workplaces, and public spaces—where they live and interact (Sandler & Pezzullo, 2007). Bringing bioethics' beneficence and nonmaleficence into the social context emphasizes that defending underprivileged neighborhoods is as important as preserving wilderness. In addition, environmental virtue ethics (Sandler & Cafaro, 2005) calls for humility, ecological empathy, and compassion since they initiate ethical actions at both the individual and collective levels. Intersectional views (Gaard, 2011; Plumwood, 2002) remind us that the social world is complex and that the burdens of pollution and achieving sustainability are not evenly distributed.

Many sustainability frameworks focus on the *economic dimension* since unchecked economic growth downplays equity requirements and respect for ecological boundaries (Kates et al., 2005; Rasmussen, 1996). By applying bioethical standards—particularly those centered on justice—we can critique profit-driven systems that exploit environmental degradation while burdening marginalized communities. Environmental justice scholarship further shows that market-centric policies, when divorced from social and ecological accountability,perpetuate inequities (Sandler & Pezzullo, 2007).

On the other hand, Jonas (1984) and Potter (1971)argue for a sustainable global economy rooted in ethical obligations to future generations. Researchers such as Peet and Watts (2004) and Newell (2019) also warn that significant changes in the political and economic order will be needed to achieve such a goal.

Mechanisms of Integration

A key step toward integrating bioethics, social ethics, and sustainability is value integration, viewed as accepting justice as a shared moral commitment. Environmental justice movements have long emphasized that the equitable sharing of environmental challenges is at the heart of ecological ethics (Sandler & Pezzullo, 2007). Bioethics supplies principles for difficult choices and compromises among multiple stakeholders' interests and conflicting socioeconomic goals (Beauchamp & Childress, 2013). In addition, the capabilities approach (Nussbaum, 2006; Sen, 1999) asks us to look at what groups and communities can realistically accomplish rather than just focusing on material developments and living standards. Intersectional and ecofeminist views (Birkeland, 1993; Gaard, 2011; Plumwood, 2002; Shiva, 1989) further call for cultural differences and past injustices to be considered when building ethical frameworks.

In terms of *institutional integration*, there is a need for effective governance, which should weave ethical values—openness, participation, and equity—into public policies (Holifield et al., 2018). Institutions should bring together various types of knowledge, from scientific data to indigenous and grassroots experiences (Berkes, 1999), as well as faith-based or philosophical principles, encouraging moral commitments (Gottlieb, 2006). As the environmental justice movement shows, bringing marginalized voices to the table often results in fairer and better ecosystem decisions (Sandler & Pezzullo, 2007).

Discussion

Critics of environmental justice have shown that traditional environmentalism involves unintentionally sidelined communities already at greater risk (Sandler & Pezzullo, 2007). Integrating bioethics, social ethics, and sustainability focuses on the issues of justice and human dignity and their role in conservation policies and environmental actions. In this process, the intersectionality and capabilities approaches ensure that we are not limited to the superficial discussions of social equity but rather try to achieve an integrated and systemic approach to sustainability.

An integrated ethical framework advises policymakers to create public policies based on fairness, cultural considerations, and solid evidence from environmental science (Kates et al., 2005). These policies should balance economic development with ethical principles (Peet & Watts, 2004; Newell, 2019), whereas normative methods (Norton, 2005) help tie public policy to local contexts, moral values, and empirical evidence. They must be supported by a broad evidence base (Berkes, 1999; Holifield et al., 2018; Potter, 1988) since sustainability decisions involve a wealth of scientific, local, spiritual, and cultural information. Bioethics advocates for integrating this diverse evidence into a reflective equilibrium —a concept articulated by Rawls (1971)—which seeks coherence among ethical principles, empirical data, and intuitive moral reasoning. It can be applied to sustainable development by balancing universal ethical ideals and sustainability principles with the lived socioeconomic conditions of different cultures and environments (Beauchamp & Childress, 2013).

Conclusion

This chapter introduces an integrative framework combining bioethics, social ethics, and sustainability theory. Viewing sustainability as a moral rather than merely a technical one shows how bioethical ideas, social ethics, and sustainability research address different parts of the multiple contemporary challenges. Along with environmental justice efforts and different theoretical perspectives, we argue that equitable and long-term sustainability solutions are needed and should be aligned with the principles of social ethics and bioethics.

References

Bauman, Z. (2011). *Collateral damage: Social inequalities in a global age*. Polity Press.

Beauchamp, T. L., & Childress, J. F. (2013). *Principles of biomedical ethics* (7th ed.). Oxford University Press.

Berkes, F. (1999). *Sacred ecology: Traditional ecological knowledge and resource management*. Taylor & Francis.

Birkeland, J. (1993). Ecofeminism: Linking theory and practice. In G. Gaard (Ed.), *Ecofeminism: Women, animals, nature* (pp. 13–59). Temple University Press.

Gaard, G. (2011). Ecofeminism revisited: Rejecting essentialism and replacing species in a material feminist environmentalism. *Feminist Formations, 23*(2), 26–53. https://doi.org/10.1353/ff.2011.0017

Gottlieb, R. S. (2006). A greener faith: Religious environmentalism and our planet's future. *Oxford University Press*. https://doi.org/10.1093/oxfordhb/9780195178722.003.0022

Grunwald, A. (2016). Responsible innovation: Bringing together technology assessment, applied ethics, and STS research. *Enterprise and Work Innovation Studies, 12*, 9–31.

Holifield, R., Porter, P., & Walker, G. (2018). *The Cambridge handbook of environmental justice and sustainable development*. Cambridge University Press.

Jasanoff, S. (2004). States of knowledge: The coproduction of science and social order. *Routledge*. https://doi.org/10.4324/9780203413845

Jenkins, W. (2013). *The future of ethics: Sustainability, social justice, and religious creativity*. Georgetown University Press.

Jonas, H. (1984). *The imperative of responsibility: In search of an ethics for the technological age*. University of Chicago Press.

Kates, R. W., Parris, T. M., & Leiserowitz, A. A. (2005). What is sustainable development? *Environment: Science and Policy for Sustainable Development, 47*(3), 8–21. https://doi.org/10.1080/00139157.2005.10524444

Meadows, D. H., Randers, J., & Meadows, D. L. (2004). *The limits to growth: The 30-year update*. Chelsea Green Publishing.

Newell, P. (2019). Global green politics. *Cambridge University Press*. https://doi.org/10.1017/9781108767224

Norton, B. G. (2005). *Sustainability: A philosophy of adaptive ecosystem management*. University of Chicago Press.

Nussbaum, M. C. (2006). Frontiers of justice: Disability, nationality, species membership. *Harvard University Press*. https://doi.org/10.2307/j.ctv1c7zftw

Peet, R., & Watts, M. (Eds.). (2004). *Liberation ecologies: Environment, development, social movements*. Routledge.

Plumwood, V. (2002). *Environmental culture: The ecological crisis of reason*. Routledge.

Potter, V. R. (1971). *Bioethics: Bridge to the future*. Prentice-Hall.

Potter, V. R. (1988). *Global bioethics: Building on the Leopold legacy*. Michigan State University Press.

Purvis, B., Mao, Y., & Robinson, D. (2019). Three pillars of sustainability: In search of conceptual origins. *Sustainability Science, 14*(3), 681–695. https://doi.org/10.1007/s11625-018-0627-5

Rasmussen, L. B. (1996). *The ethics of sustainable development: Justice, capability, and quality*. Rowman & Littlefield.

Rawls, J. (1971). A theory of justice. *Harvard University Press*. https://doi.org/10.4159/9780674042605

Sandler, R. D., & Cafaro, P. (Eds.). (2005). *Environmental virtue ethics*. Rowman & Littlefield.

Sandler, R. D., & Pezzullo, P. C. (Eds.). (2007). *Environmental justice and environmentalism: Social justice challenges the environmental movement*. MIT Press.

Sen, A. (1999). *Development as freedom*. Anchor Books.

Shiva, V. (1989). *Staying alive: Women, ecology, and development*. Zed Books.

CHAPTER 3

An Era of Mass Production of Evil

Tsuyoshi Awaya

Abstract This chapter looks at the intricate nature of good and evil, explores their manifestation and proliferation in contemporary society, and presents their contrast from bioethics and traditional ethics perspectives. Although these concepts have roots in human society, their current prevalence is essentially a result of modern capitalist systems. The concept of "human evilization" is introduced to suggest that our societal frameworks, including capitalism and socialism, contribute differently to creating environments conducive to the spread of evil—capitalism—by promoting the proliferation of minor wrongdoings through insatiable desires and socialism by giving rise to significant wrongdoings through oppressive regimes. The discussion concludes with contemplating the difficulty in combating evil, considering its ties to human desires and the power dynamics that shape our world. A thorough examination, ranging from reflections to considerations within the legal and political structure, hints at the challenging nature of eliminating evil owing to its intricate connection with fundamental aspects of human existence.

T. Awaya (✉)
Department of Law, Okayama-shoka University, Okayama, Japan
e-mail: t-awaya@nifty.ne.jp

I. Pavić et al. (eds.), *Bioethics and Social Ethics in The Modern World*,
https://doi.org/10.1007/978-3-031-86418-6_3

Keywords Evil · Human evilization · Crime of inaction · Explosion of desire · Flawed species

Introduction

Virtue and vice divide the world, but the vice often carries the greater weight. This is very understandable teaching. Of course, a good–bad dichotomy is a stereotype: it is difficult to distinguish between good and evil.

Both good and evil are the subjects of ethics and bioethics. Although these two fields overlap in some areas, they are not synonymous. Bioethics, as its definition suggests, is not a branch of ethics. It is defined as the interdisciplinary study of ethical, legal, and social issues that arise in fields such as medicine, life sciences, war and terrorism.

Benevolence is one of the four basic principles of American bioethics. In other words, right and wrong concern bioethics. However, thus far, little discussion has been conducted on this topic in bioethics. While I have always advocated evidence-based bioethics (EBB), unfortunately, this article offers no evidence, as providing specific examples is challenging.

Well, what is "evil"? The concepts of good and evil are difficult to define. However, it can be stated that it is evil, and its antithesis is good; good is the antithesis of evil. It is a fact. This is the basic structure. Good and evil are label concepts that humans create; however, are these labels meaningless? I do not think so. Therefore, it is imperative to create a better world.

Before the agricultural revolution, no good or evil could be found in the human world, just like in the animal world. During the long agricultural revolution, stealing stored surplus produce, such as grains, must have been the epitome of evil. Since then, humanity has seen a history of evil; in other words, the history of evil can be considered the history of humanity.

Evil has become part of human history. Since ancient times, it has been assumed that good and evil have become concerns for people worldwide. Even in the B.C. era, Buddha, Socrates, Mencius, Xunzi, and others spoke about good and evil. The stories of Adam and Eve in the Book of Genesis of the Old Testament are similar.

Modern capitalist societies (market economic systems) are not built upon goodness or goodwill. Companies have nothing to do with goodness but rather with making money. A slogan for peace says, "We are with you," and indeed, we are with evil. This is cruel but true.

Spread of Evil

Why is evil rampant? Why does evil spread? Why does evil win? Evil causes a chain reaction, such as a nuclear reaction. Evil tends to spread. Evil drives out good, just as bad money drives out good money. As many philosophers noted, evil is rooted in human nature.

Furthermore, it usually has a strong will and energy. Therefore, evil is strong. Generally, good people are weak. There is almost no way for a kind-hearted, good person to defeat a bad person. Moreover, the "bad guys" fight; the good guys do not. Therefore, evil flourishes, prevails, and wins.

Good people are afraid of evil. School bullies, local bosses, strong company presidents, and national dictators are frightening. Therefore, as a potent form of self-protection, good people remain silent and do nothing, which is why evil flourishes. One peace dissident activist who died recently said, "*Evil wins only because good people do nothing.*" His name and source will be withheld for some reason. At one time, John Stuart Mill said, "*Bad men need nothing more to compass their ends, than that good men should look on and do nothing*" (Mill, 1867/1994). There have been many similar comments for a long time.

This is precisely the logical structure of a "crime of inaction," a term used in criminal jurisprudence referring to a crime caused by doing nothing. We can commit crimes without action. The idea that evil wins because good does nothing follows the same logic as this "crime of inaction."

As a member of a local assembly in Japan, I once made the following statement: "To ignore evil, in other words, to turn a blind eye, is, figuratively speaking, a crime of inaction" (Awaya, 2022). The mixture was stirred without a reaction. I always think like this. I also say this in my lectures with university students: "*It is worse to not say something that should be said than saying something that should not be said.*"

Do not let evil win. Solidarity on the side of good and righteousness in the world is needed. The side of good and righteousness must prevail.

Why? Because we are humans. There is no right or wrong in the animal world. Only humans have an understanding of this concept.

I remember some impressive words by Immanuel Kant (1795/1903): "*Let justice rule on earth, although all the rouges in the world should go to the bottom.*" Of course, the notion of justice differs from good or righteousness.

Mass Production of Evil

Modern evil is mass-produced, such as clothes or computers produced in a factory. Why? Socialism creates equality and laziness, whereas capitalism and wealth directly or indirectly create corrupt people (Awaya, 2023). Specifically, the market economy and competitive society encourage "Human Evilization." Incidentally, capitalism dehumanizes humans. We are inhuman animals called humans. This is ironic.

Capitalism thoroughly stimulates human desires and enslaves people to these desires. As a result, capitalism, both structurally and semi-inevitably, directly or indirectly, results in the mass production of evil. In a market economy, people forget right and wrong, and judge things on the basis of profit and loss. This is why evil is amplified.

There are three ranks of evil: major, medium, and minor. Capitalism tends to produce many intermediate and minor evils, whereas socialism produces few significant evils. Furthermore, it is well known that socialism results in the evil of tyranny.

Descendants of Evil

We, as humans, have the ability to accept evil. While this may sound extreme, it means that we are "descendants of evil." Our ancestors were generally bad people. Evil "factors," including issues at the genetic level, have accumulated over generations. This may be an issue in Social Darwinism. We may be the descendants of people who survived because of their cowardice, cunning, smallness, ability to shift responsibility, and, to a lesser extent, of those who survived with their skill. Of course, descendants of people who survived because they were fortunate are likely to exist. In any case, we are their descendants. This is why we are here now.

The closer the era is to modern times, the higher the villain index. In other words, good people will weed out, and bad people will remain.

Many of us are descendants of the latter. That is why I think we are soft about evil or relatively tolerant. This is a hypothesis, not a significant discovery.

It is easy to understand if we consider animals and plants. Humans have been able to attain their current positions because they are cunning. In relation to animals, we are "human evil." From the point of view of cows, pigs, and chickens who are born to make marbled meat, pork shabu, and fried chicken, all humans are evil. In fact, to them, we are "human demons." Incidentally, "human dignity" is a convenient concept that humans have independently invented.

Humans are too selfish, which is the essence of anthropocentrism. Consider the Old Testament as an example: after creating humans, God said, "Dominance over the fish of the sea, the birds of the air, and all the creatures that crawl on the earth." This is a statement that is too convenient for humans. According to Nietzsche, God did not create humans, but rather, humans created God. I am not sure.

Evil and Human Desire

What is the origin of evil? It is human desire. It is said that humans are "bundles of desires." Evil arises from desire. This reminds me of Pandora's box in Greek mythology. As previously mentioned, we are slaves to our desires. Capitalism stimulates human desire, which often becomes the driving force that drives us to do evil.

Desire and evil are close friends. Human desires and conspiracy characterize the world. Humans are slaves to their desires, and those in power (rulers) are captives to their desire for power and control. Individuals who idolize those in power find themselves in the same pit, whether knowingly or not. Civilization itself is a desire-fulfillment system. To offer a more graphic description, it is a match-pump augmented reproduction and fulfillment system of desires that ignites and augments desires to fill them out and then repeats the process (Awaya, 1999). Unfortunately, our ego nests at the center of that desire.

Human desire increases explosively. The twenty-first century, like the twentieth century, is an era of the "Explosion of Desire" (Awaya, 2001), of which "Human Evilization" is a direct result. As mentioned earlier, the market economy and competitive society encourage a vicious nature in humans. It is the "Human Evilization." Of course, evilization is one type of deterioration. Modern civilization, driven by human desire and fuelled by science, technology, and the market economy, has degraded humans, effectively tightening their necks.

Evil and Power

Evil controls the human mind; in fact, it exists in everyone's mind. The Bible (New Testament) says, "*For the good which I desire, I do not do; but the evil which I do not desire, that I practice.*" This world is a battlefield between good and evil, and the battle is ongoing in our minds. Humans have one body and two minds—good and evil. When we are dominated by greed for power and feelings of hatred, our consciences become paralyzed, held captive by evil. Buddha says, "*Hurry to do good; when you are slow, your mind starts to enjoy evil.*" However, most people do not seem to be in a hurry. Eradicating evil from the world will be as impossible as eradicating viruses.

As mentioned above, many kinds of evil range from small to significant. The great evils are often based on power lust—a dark desire. Fundamentally, evil and power go hand-in-hand. Power is easily privatized. It is nothing more frightening than national power. People are constantly preyed upon for power. When we give in to the evils of power, peace comes to the surface. One classic question in legal philosophy is: "*Is bad peace better than a good fight?*" There is no answer. Even if I analyze the concept, I cannot even begin to provide an answer. Legal philosophers will continue to debate this for another 2,000 years, just as they have been doing thus far.

Power is like a drug and a pleasure for those who hold it. Of course, to gain power, it is necessary to win the power struggle. Then, a conflict involving other people arises. Furthermore, those who desire power may create new conflicts to seize power. What drives this power struggle is the desire for power, control, conquest, vanity, and ambition. Based on these findings, a "chain of hatred" occurs.

Power struggles manifest in various groups, arising in settings where two or more people come together at home, at faculty meetings, at work, with political parties, or with governments. There are power struggles among religious communities, including notable power struggles among Buddhists in Japan. Many people want to be on top. "Me and me" syndrome.

Conclusion

There was once a well-known advertising slogan for deodorants: "*You have to eliminate bad smells from the source.*" Can the roots of evil be eradicated from the source? As I have already mentioned, evil is rooted in human nature. We live in a world where evil harnesses the power of good. Humans cannot escape evil unless their brains are physically or genetically modified. Planning to eliminate the world of evil would be like turning seawater into freshwater by adding a drop of the latter.

Humans must be a "flawed species." Eradicating evil means complete human annihilation. Are we endangered species or species that are expected to become extinct? The more we know about human history, the more impressive or ugly it becomes. Disguising these features by embellishing them with fancy words is impossible.

References

Awaya, T. (1999). *Jintai buhin bijinesu [Human body parts business]*. Kodansha.

Awaya, T. (2001). *Ningen kaizo no seiki: Yokubo bigguban [Century of human remodeling: Big bang of desires]*. Shiso No Hiroba [square of Thoughts], *13*, 86.

Awaya, T. (2022). *Seiji, gyosei to seigi [Politics, administration and justice]*. Regular Assembly, Kushiro City Council, Hokkaido, Japan, Dec. 7, 2022.

Awaya, T. (2023). *Why does evil win?: The mass production of evil*, Book of Abstracts (20th International Scientific Conference of the International Society for Clinical Bioethics). Croatia, 8.

Kant, I. (1795 [1903]). *Perpetual peace: A philosophical essay*. (Smith, M. C., Trans. into English]). George Allen and Unwin, 179.

Mill, J. S. (1867 [1994]). *Inaugural address delivered to the University of St. Andrews*. Routledge and Kinokuniya, 74.

CHAPTER 4

Bioethical Principles in the Vortex of Law

Pavo Barišić

Abstract This chapter examines the emergence and establishment of bioethical principles in light of the complex and often tense relationship between law and bioethics. It explores the journey from identifying bioethical questions and problems to formulating ethical and legal regulation principles. The analysis focuses on two pivotal documents in bioethics—one in America and one in Europe. The first document is the *Belmont Report*, formally titled *Ethical Principles and Guidelines for the Protection of Human Subjects of Research.* In 1978, the National Commission for the Protection of Human Subjects of Biomedical and Behavioral Research was a foundational bioethical code for healthcare and research. The second document is the *Convention for the Protection of Human Rights and Dignity of the Human Being with regard to the Application of Biology and Medicine*, or the *Oviedo Convention* since it was opened for signature in Oviedo in 1997. This European ethical instrument plays a significant role in shaping bioethical discourse at the transnational level. Both documents focus on the role of deliberation and public debate in the formulation of fundamental ethical rules and their applications, especially

P. Barišić (✉)
Faculty of Croatian Studies, University of Zagreb, Zagreb, Croatia
e-mail: info@pavobarisic.eu

I. Pavić et al. (eds.), *Bioethics and Social Ethics in The Modern World*,
https://doi.org/10.1007/978-3-031-86418-6_4

in response to advancements in science and technology. By examining the process through which ethical principles are transformed into legal norms, this paper highlights the evolution of bioethical discourse and the essential distinction between morality and law. The analysis concludes that the interplay between law and bioethics—marked by the penetration of legal principles into bioethics and the incorporation of ethical principles into legal frameworks—is mutually enriching. Far from undermining either discipline, this integration facilitates the resolution of the profound challenges posed by modern technology and its impact on human life.

Keywords Bioethical principles · Ethical regulation · Legal regulation · Human rights · Human dignity

Introduction

The relationship between bioethics and law has been a focal point of theoretical debate for several decades. Some authors seek to protect the language of bioethics from the imposition of legal terminology. For example, one prominent legal theoretician who addressed bioethical issues and legal regulation in medicine and research accused the law of "hijacking" bioethics. Professor Emeritus Carl E. Schneider, from the University of Michigan, who passed away in May 2023, attributed the excessive influence of law to the stunting of moral reflection. He remarked on Tocqueville's phrase, "*The spirit of the law has penetrated the bosom of bioethics.*"[1] In Chapter XVI of the first volume, entitled "*Causes Mitigating Tyranny In the United States—Part I*," Tocqueville wrote:

> *The language of the law thus becomes, in some way, a vulgar tongue; the spirit of the law, which is produced in the schools and courts of justice, gradually penetrates beyond their walls into the bosom of society, where it descends to the lowest classes so that the whole people contract the habits and the tastes of the magistrate.*[2]

[1] Schneider 1994, 16.

[2] Tocqueville 1835. Cf. Graber 2004, 485 ff.

In Schneider's view, legal incursion has allegedly constrained bioethics, shifting it to a different genre of expression. However, Schneider acknowledged that legal language has, in some ways, enriched bioethical deliberation. Law offers bioethical considerations—often mired in conflicting and irreconcilable arguments—a more precise and vivid expression. Legal articulations, in turn, open pathways for resolving complex bioethical problems. In this sense, the relationship between law and bioethics could be likened to the transubstantiation of theoretical considerations into concrete social practice.

Considering the positive aspects of such *Danaan gifts* from the law, Schneider cautioned against embracing them uncritically. His caution stemmed from two key concerns. First, the goals of law differ fundamentally from those of bioethics. Second, law is inherently limited by its specific social purpose:

> *The law's gifts to bioethical discourse and effectuating that discourse should be cautiously received. For the law has goals that go beyond the immediate problems of bioethics, and those goals peculiarly shape the moral terms the law employs and specially alter the direction that legal discourse takes. Furthermore, the law has limits that arise from its special social purpose, and those limits obscure the usefulness of law's language as a vehicle for bioethical discourse.*[3]

Although the law contains richly developed conceptual tools for deliberation in many areas, Schneider considered its language inadequate for achieving the desired outcomes in bioethical matters. In support of his argument, he referred to a well-known passage from Tocqueville's *Democracy in America*, highlighting a modern tendency to translate diverse social issues into legal categories.

This chapter also builds upon Tocqueville's insightful observation of the legal normativity trend. However, a somewhat different conclusion has been reached. Rather than viewing law as merely a limiting factor, it is argued that it can also serve as a productive tool.

[3] Schneider 1994, 17. Cf. Barišić and Čović 2019, 87 ff.

Bioethics and the Spirit of Law

Moving beyond the widespread misunderstanding and mistrust between bioethics and law, it is hypothesized that ethical issues are inherently embedded in legal norms. This is particularly true for bioethical issues. The primary reason is that the modern approach to societal order is globally inclined toward legal frameworks and judicial language. What begins as ethical debate and is deliberated within ethical committees often transitions into legal norms through customary practice. It is widely acknowledged that most legal regulations originate from standards shaped by long-term use and common practices within specific local contexts. In this context, bioethics emerges as one of the most dynamic forces driving the creation of new legal norms.

The hypothesis I propose is that legal norms are not only served but also theoretically refined by bioethical considerations. This phenomenon is especially evident in biomedicine and scientific research, but it also extends to the global regulation of societal relationships and environmental behavior. There is no doubt that law draws sustenance from the principles and deliberations arising from bioethics. In contrast to the belief that this stunts ethical reflection, it is argued that it enriches and promotes it.

Examining bioethical conventions provides a lens through which to understand the ethical source of law. Since ancient times, moral issues and deliberations have gradually been codified into legal legislation—a process Tocqueville attributed to modernity but which, in reality, predates it. With the advent of democracy and the rule of law, this trend has intensified only, reflecting the organization of the state as a legal community (Barišić & Čović, 2019).

In this sense, bioethics often serves as a precursor to legal regulation. Bioethical debates allow for freer articulation of opinions, which lawyers subsequently refine into laws. On this basis, the law can be seen as "hitchhiking" on ethical developments for centuries. This chapter illustrates how bioethics exemplifies this practice by analyzing two key American and European ethical regulation documents.

The development of bioethical principles over the past half-century confirms that bioethics and law are inseparable. Moreover, bioethics should not fear legal embrace. The chapter demonstrates this by tracing the progression from everyday bioethical dilemmas to establishing universal ethical and legal principles.

The analysis focuses on two pivotal codes related to bioethics—one in America and one in Europe. The first is the *Belmont Report*, formally titled *Ethical Principles and Guidelines for the Protection of Human Subjects of Research*. This document, which was issued by the National Commission for the Protection of Human Subjects of Biomedical and Behavioral Research in 1978, outlines the fundamental ethical principles governing research involving human subjects. Given the rapid advancements in research and technology, calls for revision have emerged. For example, in reconsidering the Belmont Report, four researchers from New York University praised its significant impact and responsiveness to its time. However, they concluded that research involving human subjects has since outgrown it.[4]

The second is the *Convention for the Protection of Human Rights and Dignity of the Human Being concerning the Application of Biology and Medicine*. Known as the *Convention on Human Rights and Biomedicine* or the *Oviedo Convention*, as it was opened for signature in Oviedo in 1997, this ethical instrument has been described by Andorno as "*the first multilateral binding instrument entirely devoted to biolaw*." Andorno further highlighted its bioethical universalism, noting that "*Convention seems the best strategy for regulating biomedical research and practice at a transnational level*."[5]

Both documents focus on the role of deliberation and public debate in formulating fundamental ethical rules and their application to scientific and technological development. This discussion concerns transforming ethical principles into legal norms and examines the distinction between morality and law. It is, therefore, essential to examine the factors that have led to the adoption of new ethical principles and their integration into legal frameworks, as well as to compare the differences between approaches to bioethics in the United States and Europe.

The Belmont Report

The foundational *American Code of Bioethics* originated from the necessity of addressing doubts and problems arising from unethical research practices. One of the primary motivations was the need to establish ethical

[4] Cf. Friesen et al. 2017, 15.

[5] Andorno 2005, 143.

principles across various fields where research and experiments on humans are conducted. Legal regulation naturally evolved from these principles. Thus, as Carl E. Schneider astutely observed, contentious social and political issues are first rationalized in ethical discussions before being adopted into legal frameworks. This progression is neither exclusively modern nor uniquely American. It represents the natural course that societies have followed to structure their rules of life since ancient times.

The background of the Belmont Report is rooted in a scandal surrounding syphilis research. The United States Public Health Service and the Center for Disease Control and Prevention conducted a study involving nearly 400 African Americans infected with syphilis. This study, known as the *Tuskegee Syphilis Study* or *Tuskegee Experiment*,[6] began in 1932 in collaboration with the Tuskegee Institute, later Tuskegee University. Without the knowledge or consent of the participants, the experiment continued for 40 years, ending in 1972. It has since been labeled "*the longest nontherapeutic experiment on human beings in medical history*."[7] Public outrage over the experiment prompted President Bill Clinton to apologize to victims formally on behalf of the United States in 1997.

The *Belmont Report* came about through a uniquely American mix of guiding ethical ideals and a hands-on, pragmatic outlook. Its publication was a landmark in weaving ethical concerns into the legal fabric of policymaking. The *Belmont Report* gave legislators a clear framework for tackling multifaceted questions by boiling various bioethical issues into a few fundamental principles. This melding of ethics and law is evident in the document's wording: the law hones and expresses ethical concepts, whereas bioethical perspectives feed back into and inform legal structures.

Ultimately, the *Belmont Report* identifies three fundamental ethical principles that should be applied in research on human subjects: *respect for persons, beneficence, and justice*. These core ideas translate into guidelines, which include obtaining informed consent, ensuring the balance of risks and benefits, and fair selection of participants. Notably, *respect for persons*—and the *emphasis on informed consent*—helped reshape the

[6] Cf. About The Untreated Syphilis Study at Tuskegee, The U.S. Public Health Service, 4 September 2024.

[7] Stebbing and Bower 2004, 50.

ethical and legal landscape worldwide, elevating informed consent to a global standard in medical and scientific research.

The principle of beneficence, which emphasizes maximizing benefits while minimizing harm, reflects a distinctly American tendency to intertwine moral values with practical implications. This approach addresses ethical considerations and incorporates economic and strategic dimensions into ethical debates. Over time, legal mechanisms translated beneficence into actionable codes of conduct, thus strengthening research oversight and accountability. More recently, the idea of justice has reemerged as a fundamental ethical factor. In earlier years, several legal scholars dismissed "justice" as too theoretical to have absolute sway. The field of bioethics has reinvigorated the principle of justice and demonstrated its significant influence on ethical discourse and practical policy decisions. Indeed, justice now stands recognized as one of bioethics's four central principles, alongside autonomy, beneficence, and nonmaleficence, affirming its enduring relevance in guiding moral reasoning and legal frameworks.

The first edition of Rawls's seminal work on justice was published in 1971. In the preface to the revised edition, written in November 1990, Rawls noted that his main intention was "*to work out a conception of justice that provides a reasonably systematic alternative to utilitarianism, which in one way or another has long dominated the Anglo–Saxon tradition of political thought. The primary reason for wanting to find such an alternative is the weakness, so I think, of utilitarian doctrine as a basis for the institutions of constitutional democracy.*"[8]

The two concepts—justice and bioethics—thus became subjects of significant philosophical, political, ethical, and juridical discourse throughout the decade. Since justice resonates so strongly with legal language, highlighting the relationship between law and bioethics compels us to consider how benefits, burdens, and risks are distributed. Justice, in turn, underwrites equal treatment—a requisite for fair social relations in democratic contexts—and cements the bond between ethical reflection and legal structures.

[8] Rawls ^{5}2003, XI f.

European Convention on Bioethics

The American bioethical tradition has influenced European bioethics considerably, yet Europe's foundational code offers a distinct set of guiding principles. Formally adopted by the Council of Europe in 1997 and known as the Oviedo Convention, its full name—"*Convention for the Protection of Human Rights and Human Dignity*"—reflects a deliberate compromise.[9] Although it began as a prospective "*bioethics convention*," certain member states, most notably Germany, challenged the term "bioethics" during the 1990s and succeeded in removing it from the final title.

The process of formulating the Convention dates back to 1991 when the Parliamentary Assembly of the Council of Europe issued Recommendation 1160, titled "*Preparation of a Convention on Bioethics.*" Although disputes over terminology persisted, the Committee of Ministers approved the text in November 1996. It was then opened for signatures in Oviedo on April 4, 1997, where 35 signatories were gathered.[10] Of these, 29 ratified the Convention, embedding its principles into its domestic legislation, whereas a handful placed reservations on specific provisions. Meanwhile, the United Kingdom and Germany opted not to sign or ratify, deeming the Convention either too restrictive or too permissive.

Widely regarded as a milestone in managing bioethical concerns, the Oviedo Convention reflects the Council of Europe's goal of safeguarding human rights, democracy, and the rule of law. Legal terminology serves as the vehicle for addressing bioethical questions emerging from the dialog among 46 member states on the Council's Steering Committee on Bioethics. This framework situates human dignity at the heart of policymaking, substantially diverging from the American Belmont Report, which does not explicitly address dignity. Influenced by Kant's moral philosophy, the Convention's foremost aim is to uphold the dignity and identity of every individual, together with principles such as the

[9] The text of the Convention in several languages with accompanying explanatory notes is available on the official website of the Council of Europe at https://www.coe.int/en/web/conventions/full-list?module=treaty-detail&treatynum=164 (accessed September 27, 2024).

[10] For a detailed description of the preparations, the procedure of drafting and ratification of the document, cf. Explanatory Report. Oviedo, 4. IV. 1997.

centrality of the person, fair access to healthcare, and professional integrity in biomedical practice. These commitments appear throughout its 14 chapters and 38 articles.

Scholars, including Amir Muzur and Iva Rinčić,[11] challenge the idea that the *Oviedo Convention* mirrors American pragmatism more than older European traditions. While there is some truth in connecting it with pragmatic principles, this perspective can underplay the Convention's Kantian core. In contrast to American cost–benefit models, the European text emphasizes inherent moral values, drawing on Kant's categorical imperative. At the same time, it uses legal discussions to clarify ethical guidelines, highlighting duties and responsibilities.

Article 1 of the Convention explicitly states its intention:

> *Parties to this Convention shall protect the dignity and identity of all human beings and guarantee everyone, without discrimination, respect for their integrity and other rights and fundamental freedoms concerning the application of biology and medicine.*[12]

This clause firmly embeds respect for dignity into national laws among signatories, stipulating that each signatory nation must enact legislative measures fulfilling the Convention's provisions:

> *Each Party shall take in its internal law the necessary measures to give effect to the provisions of this Convention.*[13]

In one notable point of convergence with the American tradition, the Oviedo Convention and the Belmont Report give significant weight to consent. Nevertheless, the European framework addresses this more comprehensively, devoting Chapter II (Articles 5–9) to the nuances of consent and extending these norms to other sections on research, organ transplantation, and donation (Articles 16, 17, 19, and 20). Article 5 establishes that any health intervention must have the "*free and*

[11] Cf. Muzur and Rinčić 2017, 248.

[12] Convention for the Protection of Human Rights and Dignity of Human Being with regard to the Application of Biology and Medicine: "Convention on Human Rights and Biomedicine", Oviedo, 4. IV. 1997. Oviedo Convention, 1, p. 2.

[13] Oviedo Convention, II, 1, p. 2.

informed consent of the person concerned,"[14] requiring practitioners to offer "*appropriate information as to the purpose and nature of the intervention as well as its consequences and risks*."[15] It also affirms the right to withdraw consent freely, with supplementary safeguards for individuals who cannot consent independently.

Beyond informed consent, the Convention extensively addresses privacy rights and genetic research, providing detailed frameworks for managing new biomedical advancements more explicitly than its American equivalent. Despite these distinctions, both models underscore the intricate relationship between ethics and law, demonstrating that bioethics is fortified and structured through legal expression.

Ultimately, the Oviedo Convention illustrates a strong connection between law and bioethics. Instead of impeding ethical development, incorporating moral principles into legal frameworks enhances it and provides better responses to contemporary challenges facing biomedicine and scientific research. This illustrates how legal codification can support, rather than hinder, ethical discourse at both regional and national levels.

Conclusion

As previously discussed, the American and European bioethical conventions are closely intertwined with legal discourse, as they must address the relevant regulatory frameworks. The European Convention, in particular, draws substantially on core legal doctrines and human rights standards, making it inseparable from the legal realm. Rather than competing, bioethics and law function collaboratively.

There is a strong relationship between the development of bioethics and the law, which primarily concerns fundamental bioethical notions, including informed consent. Although informed consent is now considered a fundamental principle of medical ethics, its roots can be found in the legal discourse rather than in bioethics. Indeed, legal rulings and scholarly contributions were instrumental in formulating the notion of

[14] Oviedo Convention, II, 5, p. 2. The issue of consent is a subject of great controversy in various parts of the world. For example, Professor Tsuyoshi Awaya and a group of researchers investigated the economic and social consequences of compensated nonrelated kidney donation in the Philippines. He drew attention to possible causes of failure of informed consent (Awaya et al., 2009, 138 ff).

[15] Oviedo Convention, II, 5, p. 2.

informed consent and applying it as one of the ruling principles of bioethics. This fusion of law and ethics has broadened bioethics's scope, making it adept at navigating present-day challenges.

A significant issue arising from these observations is whether such ethical tenets can be universally applied. Foundational documents such as the Belmont Report and the Oviedo Convention seek to articulate principles suitable for all cultures and societies. Their aspiration for universality reflects Kant's moral imperatives based on generality and necessity. In this context, the legal–bioethical relationship bolsters any pursuit of principled objectivity and feasibility. More decisively than its American counterpart, the European framework demonstrates how bioethics norms can serve as a cross-border legal apparatus.

The resonance of legal tenets throughout democratic contexts, highlighted in Tocqueville's work, also appears in bioethics. Grounded in the concept of human rights, contemporary legal systems provide foundations for developing bioethical standards. In turn, bioethics refines legal reasoning by adding ethical insights that direct the pursuit of justice. As Schneider noted, the growing role of legal reasoning in bioethics has not curtailed its influence but has propelled its expansion and deepened its application.

The rapid growth of bioethics has significantly impacted the longstanding legal system. Throughout history, law and morality have operated as parallel systems for organizing society. Although moral considerations have typically preceded legal standards, their relationship has been dynamic and evolving. For the law to fulfill its primary purpose of ensuring justice, it must align with moral values.

In today's world, an integrated approach involving bioethics and law is essential for addressing global issues such as technological advancements, medical progress, climate change, and environmental sustainability. The Belmont Report and the Oviedo Convention are key examples of this integration, providing frameworks for the ethical and legal management of urgent contemporary issues.

The former President of the International Society for Clinical Bioethics, Prof Luka Tomašević, described bioethics as a "bridge of survival" and an integrative force for human thought.[16] In conclusion, I would add that in its alliance with law, bioethics not only advances

[16] Cf. Tomašević 2014, 36.

the rational organization of society but also provides a framework for navigating the complexities of modern life. Together, these fields forge pathways toward justice, equity, and the ethical stewardship of a shared human future.

References

About The Untreated Syphilis Study at Tuskegee, The U.S. Public Health Service, September 4 2024. Retrieved September 9, 2024, from https://www.cdc.gov/tuskegee/about/index.html

Andorno, R. (2005). The Oviedo convention: A European legal framework at the intersection of human rights and health law. *Journal of International Biotechnology Law, 2*(4), 133–143. Retrieved September 9, 2024, from https://www.academia.edu/457872/The_Oviedo_Convention_a_European_Legal_Framework_at_the_Intersection_of_Human_Rights_and_Health_Law

Awaya, T., Siruno, L., Toledano, S. J., Aguilar, F., Shimazono, Y., & De Castro, L. D. (2009). Failure of informed consent in compensated non-related kidney donation in the Philippines. *Asian Bioethics Review, 1*(2), 135–143. Retrieved September 27, 2024, from https://www.researchgate.net/publication/236717210_Failure_of_Informed_Consent_in_Compensated_Non-Related_Kidney_Donation_in_the_Philippines

Barišić, P., & Čović, A. (2019). Ethisierung im Licht der Dichotomie von Recht und Moral [Ethicisation in the light of law and morality dichotomy]. *Synthesis philosophica, 67*(34/1), 87–103. Retrieved September 26, 2024, from https://hrcak.srce.hr/file/326753

Explanatory Report to the Convention for the Protection of Human Rights and Dignity of the Human Being with regard to the Application of Biology and Medicine: Convention on Human Rights and Biomedicine. Oviedo, 4. IV. 1997. European Treaty Series - No. 164. Council of Europe. Retrieved September 27, 2024, from https://rm.coe.int/1680a8e4d0

Friesen, P., Kearns, L., Redman, B., & Caplan, A. L. (2017). Rethinking the belmont report?, *The American Journal of Bioethics, 17*(7), 15–21. Retrieved September 27, 2024, from https://pubmed.ncbi.nlm.nih.gov/28661753/

Graber, M. A. (2004). Resolving political questions into judicial questions: Tocqueville's Thesis Revisited. *Constitutional Commentary, 21*, 485–545. Retrieved September 27, 2024, from https://conservancy.umn.edu/server/api/core/bitstreams/91b530dc-98dc-4ab8-9b0a-f5ee8a9083e0/content

Muzur, A., & Rinčić, I. (2017). The Oviedo Convention and (European) bioethics: how much do they truly have in common? On the occasion of the 20th anniversary of the Oviedo Convention. *JAHR, 8*(2), 16, 245–250. Retrieved September 26, 2024, from https://hrcak.srce.hr/file/285964

Oviedo Convention (1997). Convention for the Protection of Human Rights and Dignity of Human Being with regard to the Application of Biology and Medicine: Convention on Human Rights and Biomedicine, *European Treaty Series – No. 164*. Oviedo, 4. IV. 1997. Council of Europe. Retrieved September 26. 2024, from https://rm.coe.int/168007cf98

Rawls, J. (2003). *A theory of justice* (Revisited). The Belknap Press of Harvard University Press.

Rinčić, I., & Muzur, A. (2010). Biomedicine and human rights: The oviedo convention and its additional protocols. *JAHR, 1*(1), 213–214; Retrieved September 26, 2024, from https://hrcak.srce.hr/file/88820

Schneider, C. E. (1994). Bioethics in the language of law. *Hastings Center Report, 24*(4), 16–22; Retrieved September 26, 2024, from https://repository.law.umich.edu/cgi/viewcontent.cgi?article=2974&context=articles

Stebbing, J., & Bower, M. (2004). Lessons for HIV from Tuskegee. *Journal of HIV Therapy, 9*(3), 50–52. Retrieved September 26, 2024, from https://pubmed.ncbi.nlm.nih.gov/15534560/#:~:text=The%20Tuskegee%20study%20of%20untreated,human%20beings%20in%20medical%20history

The Belmont Report. U.S. Department of Health and Human Services. Retrieved September 27, 2024, from https://www.hhs.gov/ohrp/regulations-and-policy/belmont-report/read-the-belmont-report/index.html

Tocqueville, A. d. (1835). *Democracy in America* (Vol. 1) (H. Reeve, Trans.). Retrieved September 26, 2024, from https://www.gutenberg.org/cache/epub/815/pg815-images.html

Tomašević, L. (2014). *Bioetika iz teološke perspektive (Bioethics from a theological perspective)*, Pergamena.

CHAPTER 5

Aristotle, Aquinas, and Integral Bioethics

Filip Veber

Abstract This chapter explores the perspectives of the "new bioethics" in dealing with the challenges presented by the philosophical perspectives of Aristotle and Aquinas. A holistic bioethics framework is proposed, transcending the theological discussions by applying the ethical sphere. Aristotle's idea of telos and its significant impacts on ethics and moral choices are explored, alongside Aquinas' reconsideration and broadening of Aristotle's teleology. The adaptation and expansion of Aristotelian teleology are analyzed, with a focus on achieving divine vision. These principles are applied to dilemmas that advocate for an approach recognizing the biological, psychological, and spiritual dimensions of human existence. The incorporation of these elements into the discussion of bioethics implies that an inclusive strategy can present empathetic answers to bioethical dilemmas. It is central to bioethics to acknowledge the concept of *natura integra*, as Aquinas advocates, promoting a unified approach to respecting the intricacies of existence.

F. Veber (✉)
Faculty of Theology, University of Ljubljana, Ljubljana, Slovenia
e-mail: filip.veber@gmail.com

I. Pavić et al. (eds.), *Bioethics and Social Ethics in The Modern World*,
https://doi.org/10.1007/978-3-031-86418-6_5

Keywords Integral bioethics · Aristotle · Thomas Aquinas · *Natura integra* · Philosophical ethics

Abbreviations

EN	*Ethica Nicomachea* of Aristotle
ENsl	Slovenian translation of Aristotles *ethica Nicomachea*
STh	*Summa theologiae* of Thomas Aquinas

Introduction

In recent years, there has been an increasing call for "new bioethics," particularly in response to particular perspectives presented by Peter Singer. This paper aims to show that the answer to this call lies in integral bioethics, rooted in the teleology of Aristotle and Aquinas, further developed through the collaboration of will and reason in human actions. As argued here, implementing integral bioethics requires a rediscovery of Aquinas' concept of *natura integra*, not solely in its theological context but also in its application to the ethical and moral domain.

Teleology in Aristotle and Aquinas

Teleology in Aristotle

The notion of *telos* is present mainly in Aristotle's *Nicomachean Ethics*, where it is explained starting from the basis that *telos* means, for the sake of which something is, to answer a simple question of "Why?" (ENsl, 387). Typically, we do not call this a cause. However, under the Aristotelian influence, the cause is usually identified with *causa finalis*, which is the ultimate reason and meaning of why we are making specific actions or acting in a certain way. *Per se*, we are used to thinking of *causa finalis* in terms of virtue or values, pushing us to follow a determined or certain modus operandi toward the specific *telos*. Aristotle is clear that only subjective, formal causes can be attributed to a conscious being, and from this, he derives the axiomatic statement that "*Omne agens agit propter finem.*" This significance will be even more significant in Aquinas. For now, let it suffice to say that it concerns this axiomatic statement that when considering free and reasonable actions, the notion of *telos* is first on the level of ideas but last on the level of action, considering that we usually follow a chain of actions before reaching the *telos*. This is important for Aristotle when he lists the finality among his four principal causes. From this perspective, the whole universe is moved toward its finality, which is, for Aristotle, resting in the unmoved mover, the uncaused cause, who gives everything the causal impulse (Stres, 2018s.v. "Smoter"). While in synthesis, this all looks fairly reasonable, the fact that, in particular, developments in philosophy, there was a strong and firm denial of teleology, which is very clear, for example, in Descartes (Mayr, 1992, p. 120) and Spinoza (Stres 2018s.v. "Smoter"), takes into consideration the following:

> *"With the rise of Galilean-Newtonian science, teleological explanations fell into general disrepute, being explicitly rejected by Bacon, Descartes, and many others of the time. Teleological explanations and scientific explanations were taken to be mutually exclusive. This extremely harsh view of teleology was combated, with only small success, by Leibniz."* (DeVries, 1991, *p. 52)*

Considering unprecedented scientific development, this decisive rejection of teleology raises questions about whether Aristotle's teleological reading still stands today.

While we could state that for Aristotle, the finality of art (*techne*) or action (*praxis*) is an a priori presupposition from the beginning (EN 1094a), on which he builds his whole ethical philosophy, we have to enquire further into this. For Aristotle, the ultimate finality has to be some "supreme good" (EN 1094a), not just any supreme good, but one that is not in knowledge but in acting, which becomes perplexing when Aristotle later defines this supreme good as happiness (EN 1095a). While at the beginning of his treaty, Aristotle does not yet distinguish between the different types of human actions and thoughts. He later states (EN 1140) that good, honestly acting (*eupraxia*) is the goal, the finality itself, which is not something extrinsic to humans but instead ordered by the virtues and intelligence (*phronesis*) that the man poses. In this sense, the quality of virtue is the quality of happiness. However, for Aristotle, the mind is above nature, which contemplates itself. According to Aristotle, this is the activity that transcends the human person, and this will bring the greatest happiness to the human person because it is the closest to divine activity (Stres, 1999, p. 108). Therefore, we see that for Aristotle, every human action and everything happening in the universe has its *telos*, which draws them toward the supreme good, toward the happiness, which lies not simply in exercising the virtues but in the persistent exercise of the virtue. Aristotle does not stop here, for he also underlines in this context the *adequation* of life with reason and intellect, which brings supreme happiness (EN 1179a).

It thus becomes apparent that the teleological underlining in Aristotle starts from happiness, which is the *telos* of human action. Nevertheless, this cannot be separated from Plato's idea of good, for only in this sense can we obtain a proper grasp and understanding. For Aristotle, *telos* is an unavoidable duty and obligation that spurs from the dynamism of life itself (Stres, 1999, p. 110). It is thus understandable that the action is not oriented toward itself but toward the ultimate *telos* of our existence (p. 27). However, the temptation of making the teleological approach a pure utilitarianism is to be repelled by understanding that the ultimate happiness does not lie in hedonism but in what was later developed by Aquinas under the notion of the blessed vision, to which both Plato and Aristotle would agree with proper understanding of it.

Teleology in Thomas Aquinas

When speaking of Aquinas, we must first consider how much Aristotelian philosophy has influenced him and to what degree Aquinas moved on and developed Aristotle under the banner of the Gospel of Christ. In this framework, we can correctly understand Aquinas when he speaks of telos, but we will only point to some of his arguments.

What is of great interest to Aquinas is that, somewhat similar to Aristotle, Aquinas does not start with whether a man has a last end if there is a *telos*, but Aquinas starts asking whether it belongs to the man to act for an end (STh I-II q. 1, a. 1).[1] In other words, it is appropriate for a human to act toward the end that he has. In what could almost be passed as a quotation from EN, Aquinas continues: *"Although the end be last in the order of execution, yet it is first in the order of the agent's intention. In addition, it is this way that is the cause"* (STh I-II q. 1, a. 1, ad 1).[2] With this, we can see that Aquinas affirms first that the end is something that

[1] *"utrum hoc sit agere propter finem" (STh I-II q. 1. pr.).*

[2] *"Ad primum ergo dicendum quod finis, etsi sit postremus in executione, est tamen primus in intentione agentis. Et hoc modo habet rationem causae."*

accords all of our actions toward itself and, therefore, cannot be the first action executed; therefore, it is called an end, goal, and not the beginning. In the same question, article 4, Aquinas rejects the objection that there is no last end of human life. He specifies:

> *"Since, therefore, good has the nature of an end, and the first good is the last end, this argument does not prove that there is no last end, but that from the end, already supposed, we may proceed downward indefinitely toward those things that are ordained to the end./../Consequently, the diffusion of goods does not proceed indefinitely, but, as it is written (Wis 11:21), God disposes of all things in number, weight, and measure." (STh I-II q. 1, a. 4, ad. 1)*[3]

Aquinas works here from the nature of the good: Any good communicates itself. We can find very similar, if not the same, principles in his deliberation on the creation: Man was created so that God could communicate His divine benefits and grace to him. Nevertheless, as Aquinas states, relying on the scripture, it pertains to God to dispose of the measure of the good communicated to someone. This remains hidden to us as part of the infinite justice and mercy of the Triune God. However, there is one good that everyone can receive in the same manner: the vision of God. Aquinas is clear: *"Final and perfect happiness can consist in nothing other than the vision of the Divine Essence"* (STh I-II q. 3, a. 8, co.).[4] For Aquinas, it is clear that the ultimate *telos* is only one. What Aristotle cannot yet articulate but calls the ultimate happiness and contemplation, which transcends human action, Aquinas can formulate clearly: the blessed vision of God, not in sacramental or any other way but *per essentiam*. As he states,

> *"Happiness is consummate perfection, which excludes every defect from the happy. In addition, therefore, whoever has happiness has it altogether unchangeably: this is done by the Divine power, which raises man to the*

[3] *"Et ideo, cum bonum habeat rationem finis, et primum bonum sit, ultimus finis, ratio ista non probat quod non sit ultimus finis; sed quod a fine primo supposito procedatur in infinitum inferius versus ea quae sunt ad finem./.../Et ideo doiffusio bonorum non procedit in infinitum, sed, sicut dicitur Sap. XI, Deus omnia disposuit in numero, pondere et mensura."*

[4] *"Respondeo dicendum quod ultima et perfecta beatitudo non potest esse nisi in visione divinae essentiae."*

> *participation of eternity which transcends all change." (STh I-II q. 4, a. 5, ad. 1)*[5]

The vision of God *per essentiam* is thus the end, the consummation and fulfillment of our happiness, perfection without any defect, never to be lost again.

However, to achieve the ultimate end, we must closely examine how Aquinas structures the way it is achieved. As stated above, we see that only so much can be done by our means. While Aquinas stands that while practicing the cardinal and theological virtues, one is acting toward his last end, we have seen that he is clear: Only God can grant and elevate us to the blessed vision. We cannot achieve this supernatural end just with our natural forces.[6] However, one still has to do what is possible to try and become worthy of receiving this grace, given to one freely by God, without any necessity on the side of God. However, Aquinas states that a human act is commanded by reason, which presupposes an act of will (STh I-II q. 17, a. 1, co.).[7] The will and reason cooperate very well in the decision-making process, as described by Aquinas. He is clear:

> *"I answer that, use of that which is directed to the end, insofar as it is in the reason referring this to the end, precedes choice, as stated above (Q16 A4)." (STh I-II q. 17, a. 3, co.)*[8]

From this quote, we see that for Aquinas in our action, two things must be observed. First, what we do is what we use is used toward its end and therefore not abused or misused. Second, even before acting, it is the reason that acting in a certain way is indeed in accordance with the end we perceive. Furthermore, he refers back to q. 16, a. 4, where he explains that the will becomes entirely proportionate to the end when it will the means in its fullness. If, then, following Aquinas, the use follows the choice, this

[5] *"Ad primum ergo dicendum quod beatitude est perfectio consumata, quae omnem defectum excludit a beato. Et ideo absque mutabilitate advenit eam habenti, faciente hoc virtute divina, quae hominem sublevat in participationem aeternitatis transcendentis omnem mutationem."*

[6] *See also STh I-II q. 5, a. 5.*

[7] *"Respondeo dicendum quod imperare est actus rationis, praesuppositio tamen actu voluntatis."*

[8] *"Respondeo dicendum quod usus eius quod est ad finem, secundum quod est in rationem referente ipsum in finem, praecedit electionem, ut supra dictum est."*

means that abuse is an act that comes from wrongly willing an end or from an irrational choice, considering that the reason that precedes the choice is then somehow mistaken that it can achieve something, which is then truly not the end of an action or use.

Integral Bioethics Today

We must return to the original definition of bioethics, as stated by Van Potter, that the point of bioethics *"is about ethical questions related to life in general, i.e., also related to the life of other living beings that could be endangered by human technoscience"* (Stres et al., 2018). He means that if we want to speak today about integral bioethics, there are several steps one has to take.

First, we must understand that ethical questions should be discussed on the basis of reason and ethical convictions, not emotions. While it is only natural that our emotional state has a specific influence on the decision-making process, we must understand that the passions in the Aristotelian sense have to be kept in line, meaning that we must rediscover the virtue of *sphrosyne*, which Aristotle in the EN understands as standing in the middle, to control oneself in regard to sensual passions, highly connected with *enkrateia*, where we give room to the voice of reason and common sense.

Second, we must return to Aristotelian and Aquinas' understanding of *telos*. Only if we start asking ourselves about the ultimate end of the material things we have and what they pose and about us as humans can we understand the importance of a rational decision-making process in ethical and bioethical questions. Understanding *telos* will help us make decisions following this approach. This will lead us toward the reasonable usage of material possessions and rational decision-making, meaning that, knowing the ultimate end, one can use the correct means to reach this end.

Third, suppose that we wish to speak about integral bioethics today in a manner that will be convincing and make people pose fundamental existential questions to themselves. In that case, we must once again reintegrate into our discussions and consider all the components of the human person. Suppose that bioethics is truly concerned with human life in general. In that case, we are to address the questions concerning the human person in their totality and with all seriousness and concern. This leads us to the mentioned Thomistic notion of *natura integra*, which comes directly from Aquinas. *Therefore, Natura integra is an orientation for integral bioethics,* so integral bioethics can work and lead us toward the complete integration of humans: the body, soul, and spirit. It comes from the name integral bioethics, meaning that the intention is to serve the human person in its totality, not just in difficult medical conditions, but to broaden the horizons of our questions. It is hardly possible for bioethics to return human beings to the state of *natura integra*, but what can (and I would argue *must*) be done is that we need to return to the logic of the *natura integra*, meaning to integrate all components of our human existence and help and approach the human person in its totality. In the 2021–2024 period, I am participating as a theological expert in research among health workers in Slovenia on their sensitivity to the spiritual component of the people they work with. Since the research has not yet been concluded, I can only say this: There is a great need among health workers (I would dare to say not just in Slovenia but also in the majority of the countries in the so-called Western world) to address the human person in front of them in totality, not just biological health or, at best, also psychological problems. Integral bioethics can thus be understood as accepting spirituality as part of the human person and addressing it.

Conclusion

On the basis of a subjective conviction, the solution lies precisely in the fullness of the notion of *natura integra*. While it is true that Aquinas uses this notion to describe the state of the human person before the Fall, it needs to be underlined once again: It is not the duty or in the domain of integral bioethics to try to create a path or return the man to the state of *natura integra* but to rediscover the logic behind it. In this state, man was able to act by following all his needs and desires but in a much different way than we do. In this state, his reason and his will were following God's will; the desires of his body, following his reason and human will, were directing his action. The beauty of the state of *natura integra* lies, on the one hand (in theological terms), in the closeness of humanity with God and in the intimate relationship that humans had with God in that state before the Fall.

On the other hand, the beauty of the state of *natura integra* also lies in the complete integration of all the components that constitute us as human beings and our lives in complete accordance with these constituents. If we manage today to comprehend this and at least try to see the human being as a whole and approach the human being with the understanding that all of his constituents are interrelated and, therefore, must be taken into account, then very much can be achieved. Nevertheless, reason and intellect must not be excluded but rather put to good use. Knowing and willing our proper end, understanding that it is beyond our grasp, yet working to be worthy of it, we can make this world better.

References

Aquinas, T. (2012). *Summa theologiae* (Vol. 13–22). Thomistic Institute.

Aristoteles. (1984). Nichomachean ethics. In J. Barnes (Ed.), *The complete works of Aristotle* (Vol. 2, pp. 1729–1867). Princeton University Press.

Aristotle, & Gantar, K. (1994). *Nikomahova etika [Nichomachean ethics]*. Trans. Kajetan Gantar. Slovenska matica.

DeVries, A. W. (1991). The dialectic of teleology. *Philosophical Topics, 19*(2), 51–70.

Mayr, E. (1992). The idea of teleology. *Journal of the History of Ideas, 53*(1), 117–135.

Stres, A. (1999). *Etika ali filozofija morale* [Ethics or Moral Philosophy]. Družina.

Stres, A., Kos, J., Petkovšek, R., & Žalec, B. (2018). *Leksikon filozofije [Lexicon of Philosophy]*. Celjska Mohorjeva družba.

CHAPTER 6

Edmund Pellegrino's Proposition of Patient-Centered Virtue Ethics

Szczepan Kaleciak

Abstract This article aims to show how Edmund Pellegrino understood the philosophy of medicine and to demonstrate how ethics based on virtue is an adequate response to such an understanding. The American bioethicist claimed that the primary goal of medicine is the patient's good, and the doctor's mission is to serve that good. Such a statement is very often misinterpreted and leads to confusion. For this reason, the author of this article will try to defend against this claim. To do that, after a brief biographical introduction, the author will discuss the sources of Edmund Pellegrino's virtue ethics and show how four types of the good of the patient can be understood: the medical good, the good experienced by the patient, the good of a patient as a person, and the patient's spiritual good. This clarification will prove necessary to properly understand the initial thesis that the goal of the medical profession is the patient's good. Only after this explanation will it become possible to show how virtues are an adequate response to seeking the good—the goal of the medical

S. Kaleciak (✉)
Pontifical University of John Paul II in Krakow, Kraków, Poland
e-mail: szczepankaleciak@pm.me

I. Pavić et al. (eds.), *Bioethics and Social Ethics in The Modern World*,
https://doi.org/10.1007/978-3-031-86418-6_6

profession. Among several virtues, Pellegrino considered the benefit of trust to be the most important. An interesting question that emerges from this reflection is related to the relationship between the types of patient good. Are any of the four more important than the others? Can any of them be sacrificed to preserve others? The author will attempt to prove that without an answer to these doubts, constructing an ethics that seeks the patient's good is impossible.

Keywords Edmund Pellegrino · Medical ethics · Virtue ethics · Patient's good · Sources of ethics · Goal of medicine

Introduction

Nearly fifty years ago, a revolution began in Western moral philosophy: the renaissance of virtue ethics. One of its most prominent representatives is the American philosopher Alasdair MacIntyre. His book *After Virtue*, published in 1981, contributed significantly to the refreshing and development of the concept of virtue in ethics, which also significantly impacted medical ethics.

MacIntyre's virtue ethics provided the basis for American physician Edmund D. Pellegrino's medical ethics. Pellegrino has written a series of books and articles in which he developed an entirely new ethical system. In his book *The Virtues in Medical Practice*, published in 1993, he presents a comprehensive philosophical theory of virtue. He applies it to the reality he knows from his daily clinical work in a hospital ward. Thus, he identifies the virtues of medical activity (Pellegrino et al., 1996). His works provide commentary and development on ethics, in which virtue is fundamental. His medical arethology is presented in works such as *For the Patient's Good: The Restoration of Beneficence in Health Care* (1988), *Helping and Healing: Religious Commitments in Health Care* (1997), *A Philosophy of Medicine Reborn: A Pellegrino Reader* (2008), or *Human Dignity and Bioethics* (2009).

This article aims to show how Edmund Pellegrino understood the philosophy of medicine and to demonstrate how ethics based on the concept of virtue is an adequate response to such an understanding. For this purpose, the first section presents MacIntyre's concept of practice and the goods internal and external to that practice. This is the foundational notion of Pellegrino's philosophy of medicine. The second part introduces the four dimensions of patient good that the author distinguishes. The final section examines the virtues Pellegrino enumerates as most relevant to medicine. These are the most adequate responses for patients to achieve good results.

Edmund D. Pellegrino's Philosophy of Medicine

The concept of practice is a pivotal notion in Pellegrino's system of ethics. He adopted this concept from the American philosopher Alasdair MacIntyre (Biesaga, 2003). MacIntyre's definition of "practice" does not always align with its common understanding. Therefore, it is crucial to explain this term before discussing the notion of virtue (MacIntyre, 2007).

Practice is defined as any complex, coherent, and socially established form of cooperative activity during which the intrinsic goods of that activity are realized (MacIntyre, 2007, s. 187). These goods are realized when the patterns that define a particular activity are followed (Knight, 2008). The pattern of activity is nothing more than what is most characteristic of it—it is an ideal model. The set of practices understood in this way includes activities such as chess, architecture, leadership, and medicine (MacIntyre, 2007, s. 187–188).

The indispensable elements of practice are the patterns of excellence achieved through obedience to certain preestablished principles. Entering into practice requires conforming to these patterns and acknowledging the existing authorities that define them (Wierzejski, 2014).

The analysis of the aforementioned definition reveals that two types of goods are achievable through the exercise of practice: external and internal (Knight, 2013). External goods are only incidentally associated with it and do not belong to its essence or directly participate in achieving practice-specific objectives. There are various alternative ways to obtain them. These could be, for example, money earned, fame gained, power, or social privileges. Once they are achieved, they become the property of the individual. Furthermore, the more one has, the less available for others, making them an object of competition (Gałkowski, 2004). In the case of the therapeutic relationship, both parties are antagonized because of them (Biesaga, 2004).

Internal goods are substantially linked to practice. They can only be achieved through participation in practice by performing specific acts (Higgins, 2010). Describing them is only possible through concepts and examples from the given practice (Pellegrino, 2001). Attaining them benefits all parties involved (MacIntyre, 2007). For this reason, unlike those discussed above, they do not become objects of competition. Internal goods are nonconflicting and enrich both parties involved in the therapeutic relationship, namely, the patient and the practitioner, and consequently, the whole community to which they belong (Biesaga, 2004).

There are many types of practices, such as arts, crafts, and science, so they are oriented to various goods. As the goal of the art of shipbuilding is to build ships, the goal of management is prosperity, so the goal of the medical profession is health. As Pellegrino noted, since the medical profession fulfills the criteria proposed by MacIntyre in his definition of practice, it can be analyzed in terms of internal and external goods. Excellence in applying treatment standards or patient health are internal goods, and making money or gaining a reputation are external goods to that practice (Pellegrino, 2001, s. 562).

A physician may seek various external goals related to his profession, and it is neither wrong nor dishonorable that he becomes respected, enriches himself, or gains fame through his work. However, distinguishing between these and the internal goal of medicine, which is the patient's good, is important.

Dimensions of a Patient's Good

Pellegrino proposed a medical ethics centered around the concept of virtue, which, when followed, leads to good (Pellegrino, 2000). The goal of medical practice is the internal good, which can be defined as the good of the patient. Pellegrino's theory rejects utilitarian or consequentialist approaches that involve social or economic cost/benefit calculations. The internal goals of medicine are not the same as those imposed on it by the state or society. Pursuing only external goals can lead to a deviation from the specific purpose of the practice, resulting in a loss of its essence (Biesaga, 2004).

According to Pellegrino, the ultimate goal of medicine is to restore and develop the patient's health, that is, to heal him or her and cure his or her disease and illness (Markee, 2017). When this is not possible, the goal is to ensure that the patient lives his life to the fullest despite any discomfort, pain, or disability (Pellegrino & Thomasma, 1993). This commitment to medicine is nonnegotiable and unquestionable. It grows out of the nature of the doctor–patient relationship and belongs to the essence of the medical profession (Vizcarrondo, 2012). Medicine addresses patients who expect to be cured. Practitioners are in a relationship with someone in need (Pellegrino, 2001). This is the most primordial situation for the doctor–patient relationship, becoming the *locus ethicus* for medical ethics (Pellegrino, 1978).

Pellegrino distinguished four components of the broader concept of the patient's good: the medical good, which is determined by manuals and guidelines; the good, which is experienced by the patient following his or her more or less adequate assessment of his or her situation, value system, and future life plans; the good of the patient as a person; and the spiritual good of the patient (Pellegrino, 1988).

The first is the medical good of the patient, which is determined by current scientific medical knowledge. It is objectively expressed in the current standards and guidelines of conduct. They allow medical practitioners to use adequate means to prevent, diagnose, and treat the disease. This good aspect includes restoring or improving physical and mental functioning, relieving pain and discomfort, or improving physical functioning (Biesaga, 2004). This aspect of patient good is essential and fundamental to medicine, but it is insufficient. Medicine may not end here. In contrast, as the starting point, the medical good must be harmonized with the subsequent dimensions of the broader concept of the patient's good.

Pellegrino identifies the patient's concept of his good as the second dimension of the patient's good (Wasson & Cook, 2006). This refers to the patient's personal preferences and choices that result from the world of values he affirms. The patient can decide how he or she wants to live his or her life and what level of inconvenience of the proposed therapy he or she is willing to accept. Age, experience, social and religious involvement, and relationships with relatives who support the patient through the disease are important in identifying what constitutes the patient's good (Biesaga, 2004). State-of-the-art diagnosis and treatment techniques may be entirely useless for a patient who does not accept the illness.

The third dimension of the patient's good, as mentioned by Pellegrino, is the good of the patient as a person. The philosopher argues that it is based on the inalienable, immutable personal dignity that is inherent in the ontological structure of every human being. Personal dignity means that an individual's value does not depend on health, gender, education, ethnic background, or social position. Being a person is good in itself. This good must not be used as a means to any end, even the most noble one (Biesaga, 2016). Pellegrino refers to the so-called personalist norm (Wojtyła, 1982). Classical and personalist ethics, to which Pellegrino alludes, assume that personal dignity does not depend on customs or the social contract but is grounded in man's nature.

The concept of a patient's spiritual good is understood more broadly than religiously, although it may be related to the patient's religion. This includes all aspects that give life its ultimate meaning and orient toward its ultimate destiny, such as faith, religious beliefs, or a value system. For them, man is ready to make the most significant sacrifices (Biesaga, 2004).

Pellegrino argues that there is a hierarchy among the four dimensions of the patient's good. The spiritual good is situated at the highest level, followed by the good of the patient as a person and then the good that the patient experiences. At the lowest level is the medical good, understood as the technical realization of current medical feasibilities.

It is essential to consider its four components to achieve holistic patient good. Therefore, medical practitioners require proficiencies to ensure proper attitudes in their relationships with patients. While medical knowledge and technical skills are necessary for realizing the medical good of the patient, they are not sufficient. Acquiring appropriate character skills, or virtues, is even more crucial.

Virtue as an Adequate Response to Seeking the Patient's Good

Pellegrino's concept of medical practice and the hierarchy of patients' goods constitute a starting point for his theory of medical virtue, an essential part of his philosophy of medicine. His theory defines virtue as an acquired property that enables a person to acquire goods internal to a given practice, in this case, a medical one. Without them, the attainment of these goods is effectively prevented. MacIntyre understood virtue as a disposition or acquired quality that is necessary (1) to achieve the good internal to practice; (2) to sustain communities in which the individual can seek the highest good; and (3) to sustain tradition, which creates the necessary historical context for the life of each individual and community (MacIntyre, 1988, 74).

According to the theories of Aristotle and MacIntyre, Pellegrino defines virtue as a character aptitude that habitually disposes the possessor to have the right intention and to act with respect to the purpose of medical practice. Virtue empowers reason and allows one to perceive and adopt motivation, aiming ideally at a moral goal (Pellegrino, 1995, 267).

Pellegrino defines virtues as related to the goal of medicine. These include fidelity to trust and promise, benevolence, effacement of self-interest, compassion and caring, intellectual honesty, justice, and prudence (Pellegrino, 1995). The virtues are described in the doctor–patient relationship context, where trust is essential. The doctor is responsible for the patient (Pellegrino, 1990, 226–227). This relationship is not symmetrical since the patient does not choose his condition and, therefore, has no choice but to trust the medical practitioner (Pellegrino, 2005, 477–478). The physician is responsible for not violating this trust (Pellegrino & Thomasma, 1993). Benevolence should be realized on this basis. Diverse benefits could result from misusing the patient's vulnerability, which is why a physician must be able to put aside his interests. In this context, the fundamental role of justice should be understood. Pellegrino's subsequent works have evolved a catalog of virtues to include compassion, phronesis, justice, fortitude, temperance, integrity, and self-effacement (Pellegrino & Thomasma, 1993). However, despite slight differences, the presented ethics remain focused on the fundamental principle of beneficence in trust (Pellegrino, 1988).

Trust is a fundamental aspect of human relationships that cannot be excluded from ethical discourse. This moral obligation is inherent to medicine and is due to the nature of this profession (Pellegrino, 1983). Pellegrino's medical ethics consider benevolence in trust to be the cardinal principle. All human relationships involve mutual trust, as does the doctor–patient relationship, which has an additional moral dimension (Ayibo, 2020). The patient begins to seek help when he notices a symptom that risks his well-being and health. Seeking help from a healthcare specialist implies that the specialist has the knowledge and abilities necessary to aid. Trust is assumed from the beginning of the patient's actions. He trusts in the existence of a therapy that can help him, in the fact that the doctor is competent, and in the goodwill of medical personnel (Baier, 1986). From the beginning, the therapeutic relationship is not equal, as the medical specialist holds the knowledge and authority. At the same time, the patient carries the burden of uncertainty, anxiety, pain, suffering, and fears (Pellegrino, 1987, 11–12).

Pellegrino faced many critical arguments. One of them was formulated by Veatch, whose contractualism presumes that trust is not possible between members of society (Veatch, 2006). Pellegrino calls this "the ethics of distrust." Veatch's main argument for the impossibility of mutual trust is an example of the many deviations and pathologies that occur because of physicians. According to him, experience shows that medical professionals can sometimes be incompetent, dishonest, lie, and addicted to alcohol or other psychoactive substances (Veatch, 1991). This can make it difficult to trust them. The paradigm of their work is not the highest good of the patient, who can become merely a means to achieve personal goals or even an obstacle to achieving them. According to Veach, the remedy for deviations should be the implementation of licensing, certification procedures, quality controls, periodic renewal of diplomas, or strict legal responsibility. All these propositions demonstrate a lack of trust in medical professionals. Pellegrino replies that every professional is, first and foremost, a human being and that human nature is burdened with a propensity for evil. A certain degree of distrust caused by the unpredictability of another's behavior is unavoidable. However, this does not justify an ethic rejecting trust in the doctor–patient relationship (Pellegrino & Thomasma, 1993).

Trust must be merited by years of practice, gaining knowledge and experience, and one's attitude and behavior. No one can claim trust solely on the basis of his title or education. Trust cannot be removed from a relationship. It is a moral obligation, not a privilege. More attention is required to form the character and professionalism of young medical professionals since virtue is gained through practice and is best learned from those masters who set a good example (Agich, 1980). The alternative conception of medicine is one in which the main preoccupation is patients' self-defense against potential error and exploitation and doctors' self-defense against lawsuits and legal consequences. A specialist without the moral obligation of being trusted will be free to abuse his patients, seeking his benefit.

Conclusion

Pellegrino formulated a comprehensive philosophy of medicine and analyzed its essential facts: illness, the act of medicine, and the act of treatment. He then evoked the *telos* of this specific activity and framed it in terms of the four goods that make up the patient's good. Finally, he distinguished between goods internal and external to medical practice.

Trust is the element that gives the right direction to medical ethics. Pellegrino's theory opposes popular currents of modern medical ethics, such as utilitarianism, principlism, or contractualism. These currents face many difficulties when confronting the distinction between internal and external goods, which requires further in-depth research.

Virtue is defined in relation to medical practice. This does not imply the approval of all practices under all possible circumstances. Further research is needed to explore how MacIntyre's concept, developed by Pellegrino, can be applied to other medical professions, such as nursing, obstetrics, physiotherapy, or medical diagnostics. Furthermore, the morality of virtue characteristic of the medical profession needs to be complemented by the concept of moral law, which the practices must also adhere to. The next challenge for contemporary moral philosophy is to examine the relationship between virtues and moral law, that is, between what is right and wrong in an objective way.

References

Agich, G. J. (1980). Professionalism and ethics in health care. *Journal of Medicine and Philosophy, 5*(3), 186–199. https://doi.org/10.1093/jmp/5.3.186

Ayibo, P. (2020). A virtue-based model for medical ethics and practice in Edmund Pellegrino. *Philosophy (0861–6302), 29*(1), 20–30.

Baier, A. (1986). Trust and antitrust. *Ethics, 96*(2), 231–260. https://doi.org/10.1086/292745

Biesaga, T. (2003). Edmund D. Pellegrino's concept of the aim of medicine. W T. Biesaga (Red.), *Systemy bioetyki* (pp. 9–22). Wydawnictwo Naukowe PAT.

Biesaga, T. (2004). Patient's good as the goal of medicine and basis of medical ethics. *Studia Philosophiae Christianae, 40*(1), 153–165.

Biesaga, T. (2016). The dignity of human person. A source of the personalistic norm. W. T. Duma, A. Maryniarczyk, & P. Sulenta (Eds.), *Wokół antropologii Karola Wojtyły* (s. 219–236). Polskie Towarzystwo Tomasza z Akwinu.

Gałkowski, S. (2004). Virtues and relativism. Alisdair MacIntyre's attempt to overcome relativism. *Diametros, 2*, 1–17. https://doi.org/10.13153/diam.2.2004.43

Higgins, C. (2010). Worlds of practice: MacIntyre's challenge to applied ethics. *Journal of Philosophy of Education, 44*(2/3), 237–273. https://doi.org/10.1111/j.1467-9752.2010.00755.x

Knight, K. (2008). Practices: The aristotelian concept. *Analyze & Kritik, 30*(1), 317–329. https://doi.org/10.1515/auk-2008-0118

Knight, K. (2013). An argument within aristotelianism: Maritain and MacIntyre on the theory and practice of human rights. *Hungarian Philosophical Review, 57*(4), 72–85.

MacIntyre, A. C. (1988). *Whose justice? Which rationality?* University of Notre Dame Press.
MacIntyre, A. C. (2007). *After virtue: A study in moral theory* (3. ed). University of Notre Dame Press.
Markee, M. D. (2017). The aristotelian ethics of collaborative care. *Journal of Interprofessional Education & Practice, 9*, 115–120. https://doi.org/10.1016/j.xjep.2017.09.005
Pellegrino, E. D. (1978). Ethics and the moral center of the medical enterprise. *Bulletin of the New York Academy of Medicine, 54*(7), 625–640.
Pellegrino, E. D. (1983). The healing relationship: The architectonics of clinical medicine. W E. E. Shelp (Red.), *The clinical encounter: The moral fabric of the patient–physician relationship* (s. 153–172). Springer Netherlands. https://doi.org/10.1007/978-94-009-7148-6_11
Pellegrino, E. D. (1987). Toward a reconstruction of medical morality. *The Journal of Medical Humanities and Bioethics, 8*(1), 7–18. https://doi.org/10.1007/BF01119343
Pellegrino, E. D. (1988). *For the patient's good: The restoration of beneficence in health care* (D. C. Thomasma, Red.). Oxford University Press.
Pellegrino, E. D. (1990). The medical profession as a moral community. *Bulletin of the New York Academy of Medicine, 66*(3), 221–232.
Pellegrino, E. D. (1995). Toward a virtue-based normative ethics for the health professions. *Kennedy Institute of Ethics Journal, 5*(3), 253–277. https://doi.org/10.1353/ken.0.0044
Pellegrino, E. D. (2000). Bioethics at century's turn: Can normative ethics be retrieved? *The Journal of Medicine and Philosophy, 25*(6), 655–675. https://doi.org/10.1076/jmep.25.6.655.6133
Pellegrino, E. D. (2001). The internal morality of clinical medicine: A Paradigm for the ethics of the helping and healing professions. *The Journal of Medicine and Philosophy, 26*(6), 559–579. https://doi.org/10.1076/jmep.26.6.559.2998
Pellegrino, E. D. (2005). Some things ought never be done: Moral absolutes in clinical ethics. *Theoretical Medicine and Bioethics*, *26*(6), 469–486. https://doi.org/10.1007/s11017-005-2201-2
Pellegrino, E. D., & Thomasma, D. C. (1993). *The virtues in medical practice.* Oxford University Press. https://doi.org/10.1093/oso/9780195082890.001.0001
Pellegrino, E. D., Thomasma, D. C., & Miller, D. G. (1996). *The Christian virtues in medical practice.* Georgetown University Press.
Veatch, R. M. (1991). Is Trust of professionals a coherent concept? In W. E. D. Pellegrino, R. M. Veatch, & J. Langan (Eds.), *Ethics, trust, and the professions: Philosophical and cultural aspects* (139–158). Georgetown University Press.

Veatch, R. M. (2006). Assessing Pellegrino's reconstruction of medical morality. *The American Journal of Bioethics*, *6*(2), 72–75. https://doi.org/10.1080/15265160500507082

Vizcarrondo, F. E. (2012). The return of virtue to ethical medical decision making. *The Linacre Quarterly*, *79*(1), 73–80. https://doi.org/10.1179/002436312803571519

Wasson, K., & Cook, E. D. (2006). Pellegrino and Medicine: a critical revision. *The American Journal of Bioethics*, *6*(2), 90–91. https://doi.org/10.1080/15265160500507132

Wierzejski, W. (2014). Virtue ethics in the context of the overall concept of Alasdair MacIntyre's moral inquiry. *Studia Philosophiae Christianae*, *50*(3), 57–80.

Wojtyła, K. (1982). *Miłość i odpowiedzialność*. Towarzystwo Naukowe KUL.

CHAPTER 7

Ethical and Legal Dilemmas of Surrogacy: *Gain for One, Loss for None?*

Nevena Aljinović and Ana Jeličić

Abstract Scientific and biomedical achievements have enabled infertile couples to achieve parenthood through assisted fertilization methods as an alternative form of reproduction. Surrogacy is the best example of the far-reaching nature of assisted reproduction and represents the most complex medical-legal-ethical form of human conception. It polarizes public opinion and encourages moral, ethical, and legal debates. It calls into question the traditional concept of the family, especially motherhood, because it contradicts the social and legal presumption that the child's mother is the woman who gave birth (*mater semper certa est*). Recognition of the citizenship of children born by international surrogacy has become a burning legal issue in the focus of international debate. The European Court of Human Rights (ECtHR) also addresses the above problem, which places the child's best interest on the pedestal of social

N. Aljinović · A. Jeličić (✉)
University of Split, University Department of Forensic Sciences, Split, Croatia
e-mail: ajelicic@forenzika.unist.hr

N. Aljinović
e-mail: nevena.aljinovic@forenzika.unist.hr

I. Pavić et al. (eds.), *Bioethics and Social Ethics in The Modern World*,
https://doi.org/10.1007/978-3-031-86418-6_7

priorities and violates the child's rights. Ethical dilemmas raised by surrogacy concern preserving human dignity, exploiting surrogate mothers, and engaging in child trafficking. For this reason, many countries prohibit surrogacy. However, despite the indisputable complex legal and ethical dilemmas, surrogate motherhood is gaining increasing legitimacy in many countries worldwide. This is evident from the implicit acceptance and court approval of surrogacy in numerous American states. Additionally, some countries legally regulate surrogacy, prescribing (strict) conditions that must be met by both contracting parties to fulfill "contractual obligations successfully." Finally, the latest trends are toward adopting international normative and ethical guidelines for protecting the rights of children born through international surrogacy.

Keywords Surrogacy · Surrogate motherhood · Motherhood · Parenting · Bioethics · Law

Introduction to Surrogacy

A surrogate is a woman who agrees to conceive and give birth to a child to give it to another person or couple who acquires parental rights and obligations over the child (Patel et al., 2018, p. 212; Radan et al., 2015, p. 37; Ashley, 2006, p. 88). There are numerous modalities of performing surrogacy, but the most common is the division into genetic (traditional) or partial and gestational or complete surrogacy (Margalit, 2014; Pande, 2010; Spivack, 2010). The difference between these two forms lies in the (non)existence of a genetic link between the surrogate and the child she is carrying. Thus, in genetic (partial) surrogacy, there is a biological connection between the surrogate and the child since the ovum of the surrogate is fertilized with the reproductive cell (sperm) of the intended father, either in vitro or, traditionally, by sex. The genetic surrogate is the baby's biological mother. In gestational (full) surrogacy, there is no biological connection between the surrogate and the child she is carrying since the embryo is created from the gametes of the intended parents (rarely the donor), which are fertilized in vitro and implanted in the surrogate (Caster, 2010; Ryznar, 2009; Spivack, 2010). Depending on whether the surrogate receives special financial compensation (or not) for the "service" provided (in addition to compensation for costs caused by pregnancy and childbirth and legal expenses), we are talking about a commercial or an altruistic form.

Human reproduction interferes with social, political, and religious aspects and has moral (bio), ethical, and legal repercussions. Surrogacy challenges the nexus of "woman" and "mother" as something unrelated to pregnancy. The conflicting social view of the justification of surrogacy and its (moral) acceptability causes numerous (bio)ethical questions, split social reactions, and an uneven legislative approach to surrogacy regulation, oscillating from a complete ban to detailed legal regulation. Many European countries expressly prohibit surrogacy, such as Croatia (Medically Assisted Reproduction Act, OG 86/12, art. 31), Germany (The Embryo Protection Act—ESchG, art. 1.), Italy (Law No. 40 of 19 February 2004 on Medically Assisted Procreation, art. 12, paragraph 6), Spain (Law 14/2006, of 26 May, on Assisted Human Reproduction Techniques, art. 10) or Switzerland (Reproductive Medicine Act—FMedG, art. 4), and some go so far as to consider it a criminal offense (e.g., French Penal Code, art. 227-13). Some states only accept the altruistic model, such as all Australian states and territories (but eligibility requirements may vary; see: Surrogacy Act, 2010; Parentage Act, 2004; Family Relationships Act, 1975), Canada (Assisted Human Reproduction Act, S. C. 2004, c. 2), or Israel (The Embryo Carrying Agreements Law 5756-1996). Other countries, such as Russia (Family Code of Russia, art. 51) or Ukraine (Family Code of Ukraine, art. 123), allow both commercial and altruistic models (González, 2020; Margalit, 2014).

In countries that allow surrogacy, it can either be regulated by a special law (e.g., in some American states), or the practice of surrogacy is carried out without any special law according to the principle that "*whatever is not prohibited is allowed*" (e.g., in Belgium and the Netherlands) (González, 2020, p. 917). Furthermore, countries that allow surrogacy may have a closed or open approach, meaning that surrogacy is limited exclusively to its nationals (e.g., India, Israel, or Thailand), or it may be available to everyone (e.g., Ukraine) (González, 2020). However, even when domestic surrogacy is legal, the intended parents may opt for international surrogacy for reasons of an economic nature (lower costs) or noneconomic reasons (e.g., because, as gay couples in their home country, they are not eligible candidates for surrogacy or artificial insemination) (Inhorn and Patrizio, 2009).

In this paper, we are not concerned primarily with presenting medical methods of surrogacy but rather with analyzing the legal implications of the lack of uniform law regulation and highlighting the (bio) ethical problems implied by this method of parenthood.

Key Legal Issues Caused by (International) Surrogacy

International surrogacy brought up the issue of recognizing the citizenship of children born abroad, and the Baby Manji case (Baby Manji Yamada v. Union of India, (2008) 13 S.C.C. 518) raised it to the pedestal of heated debate. In the context of the above, such children may de facto become stateless owing to the nonrecognition of the citizenship of the country in which they were born via surrogates and the denial of the acquisition of citizenship of the country from which the intended parents are citizens. Further problems that may arise are the impossibility of obtaining a travel document for a child who has not yet acquired citizenship (e.g., the Balaz case) or the impossibility of registering the intended parents in the birth register as child parents (cases of Mennesson v. France, application no. 65192/11, Judgment of 26 June 2014 and Labassee v. France, Application No. 65941/11, Judgment of 26 June 2014).

Because of the aforementioned undesirable byproducts of international surrogacy, undeniable efforts have been made at the international level "*to identify the most problematic areas and to formulate procedural and safeguard requirements to ensure the protection of the rights of children born through surrogacy*" (Verona Principles, 2021, p. 3). Therefore, at the beginning of 2021, independent international law and human rights experts drafted the "*Verona Principles for the Protection of the Rights of a Child Born Through Surrogacy*," which aims to stimulate discussions and changes at the legislative, political, and social levels. Although the principles are not legally binding, they represent an important contribution to the creation of normative guidelines for the protection of the rights of children born through surrogacy. However, the adoption of the Verona Principles also triggered an avalanche of criticism that "*under an alleged concern for the rights of children born through surrogacy, it aims at regulating cross-border surrogacy*" (CIAMS, 2022). Opponents mainly state that special attention should be given to two normative initiatives. The first is a "*protocol that will fuel the worldwide legalization of surrogate practice*" (brought by the Hague Conference on Private International Law—HCCH), and the second is the principle that "*will have the effect of encouraging states to facilitate cross-border surrogacy*" (brought by the International Nongovernmental Organization—ISS) (CIAMS, 2022).

The European Court of Human Rights (ECtHR) also dealt with the recognition of the citizenship of children born through surrogacy in the context of a violation of Art. 8. The European Convention for the Protection of Human Rights and Fundamental Freedoms (ECHR) protects the right to respect for private and family life. Two important decisions were made beginning in 2014. (Mennesson v. France 2014 and Labassee v. France 2014), the ECtHR, guided by the best interests of the child, found a violation of the child's right to privacy owing to the nonrecognition of the parental status of the intended parents and children born abroad by the surrogate and, consequently, the rights of children to acquire citizenship (of their parents). The ECtHR considered denying parental status, despite having an undoubted genetic link with one of the intended parents, "*undermined the children's identity*" (Mennesson § 96). Although Art. 8 of the ECtHR does not guarantee the right to acquire (specific) citizenship; citizenship is an (essential) element of a person's identity. These ECtHR decisions practically led to the revision of (European) judicial practice in the direction of recognizing parentage in situations where there is a genetic link between a child and (at least) one intended parent, which de facto led to the oxymoron that a country that explicitly prohibits the practice of domestic surrogacy must "allow" it in regard to international surrogacy (Iliadou, 2018, p. 153).

Bioethical Issues Related to Surrogate Motherhood

Moral critics of surrogacy single out its commercialization in the first place. They consider it ethically inadmissible to give financial compensation to indigent surrogates since, without monetary need, they would not voluntarily agree to this act. It is ethically unacceptable to take advantage of the poverty of a woman who decides to be a surrogate because it is her source of income. In this way, the exploitation of impecunious women by the higher and wealthier social class occurs (Macklin, 1991, p. 148).

Surrogacy is all the more ethically debatable if the woman does not have health problems but refuses pregnancy for aesthetic and other reasons but does not renounce; in contrast, she stands up for biological parenthood. In that way, women and the female body are dehumanized and turned into market goods (Roksandić Vidlička et al., 2012, p. 44). Moreover, potential surrogates are subjected to various forms of discrimination because they are chosen on the basis of their age, physical appearance, or health condition.

However, these are not the only bioethical problems related to surrogacy. We will outline several potential ethically and morally debatable scenarios. It is possible to imagine a situation where the intended parents give up parenthood and ask the surrogate to abort the child. In this case, it is a morally susceptible and debatable question of the permissibility of a surrogate being forced to have an abortion that she conscientiously opposes or may not want because she perceives it as risky for her health. In addition, surrogacy can genuinely represent a health risk for a woman because her body can recognize and then reject an implanted embryo "*with chromosomes that are completely different from her own*" as a foreign body (Roksandić Vidlička et al., 2012, p. 46). In the case of a conflict of interest between the surrogate and the child, it is necessary to be guided by a moral theory based on equality and the everyday needs of the pregnant woman and the child (Harris, 2000, p. 789). This *ethical approach* guarantees the safety of the pregnant woman and the child without encroaching on the privacy and physical integrity of the pregnant woman. It must have a solid foundation in education but also in *ethical compulsion* by which everything is done to make the pregnant woman understand that it is her moral obligation not to harm the fetus with her unhealthy habits and behavior (Frković, 2010, p. 183; Frković & Katalinić, 2006).

Potential threats to the child can be unhealthy behavior and bad lifestyle habits of a surrogate who does not feel responsibility or love for the child she is carrying and does not show excessive care, i.e., prenatal care that is typical of a biological mother. This is an important ethical issue, although the doctrine of the autonomy of pregnant women has been reduced. In many countries, especially in the United States of America, legal proceedings are taken against women who, for example, smoke or consume alcohol during pregnancy because such behavior harms the development of the fetus. The goal is not to punish the woman but to protect the unborn child (Frković, 2010, p. 182).

Furthermore, surrogacy can cause interpersonal problems between the intended parents, the surrogate, and her family, especially the surrogate's husband and her biological children. In addition, there are threats related to the relationship of the surrogate to the child with whom she can quickly bond emotionally during pregnancy, as well as the problem of a child who is born by a surrogate and later has problems defining her own identity (Roksandić Vidlička et al., 2012, p. 52).

Serious bioethical dilemmas can be caused by a situation in which it is determined by the prenatal diagnosis that the surrogate is carrying a child with developmental difficulties, which the intended parents refuse to accept (the Baby Gammy case from 2014), or the child is unexpectedly born with malformations or any other, for intended parents, unacceptable defects, difficulties, and diseases, which is why they refuse to take it over.

A bioethical problem is a possible divorce of the intended parents that occurred during the surrogate pregnancy, followed by a struggle for the child (as in the case of Baby Manji) or an even more difficult situation when neither intended parent wanted the child of their ex-spouse or did not want to be a single parent. Of course, it can also happen that intended parents obtain their biological child naturally and that the child carried by the surrogate becomes redundant or that the intended parents do not accept it in the same way as the child that the intended mother gave birth to on her own. In such cases, when deciding upon the best interests of the child, many persons in the surrogate process may remain legally and morally deprived of their rights.

Some authors, from a bioethical perspective, consider the absence of a genetic and biological link between the surrogate and the child she carries to be problematic since a natural emotional bond and closeness with the child, which is naturally created during pregnancy, is not made. It can cause psychological and emotional difficulties for a surrogate (Türk & Terzioğlu, 2014, pp. 101–102). Similarly, there is a possibility that the intended mother fails to bond with the child born by the surrogate because there is no bonding during pregnancy.

Furthermore, surrogates may suffer from emotional difficulties caused by giving birth and handing over the child to the intended parents. Emotional difficulties are certainly more significant if the surrogate is also the child's biological mother. A study conducted by Jadva et al., (2003, pp. 2200–2201) with 34 surrogates indicated that 32% of them suffered from mental and emotional difficulties after handing over the child, and 9% requested the support of a psychologist. Therefore, the question remains as to how well a surrogate can be informed and give informed consent to something for which she cannot know in advance what kind of emotional difficulties it may cause her.

It is also necessary to mention the problem related to age discrimination, that is, the question of the age limit at which people can claim and exercise their rights to parentage through a surrogate. Recently, the world media reported on a woman who had a grandchild born via surrogacy who was conceived by artificial insemination, i.e., using her deceased son's sperm (which was kept in cryopreservation) and donor ovum (Večernji.hr, 2023). In this case, it is ethically debatable that the child born will never have the opportunity to meet its biological parents, and the rest is in the care of a woman on the threshold of seventy years of age. Bioethically, the issue of the permissibility of the procreation of persons who have passed away and whose gametes were preserved and frozen during their lifetime is controversial. In this way, science and individuals, disposing of the body of a dead person, create children from deceased parents, denying them the essential right to a family and parental care. In contrast, the child becomes the object of fulfilling the selfish wishes of the deceased and living persons who do not think about the child's best interest.

In the end, we look back at the issue of so-called procreative tourism. Intended parents who do not have the possibility of using surrogacy in their countries of residence or stay (because it is prohibited) may visit countries where this practice is legal (procreative traveling), where they pay women for surrogacy services with the help of surrogacy agencies. The financial commission profit of agencies for the surrogate (*global baby business*) is also ethically very debatable, and there is the eternal problem of the risk of trafficking in women and children (Piersanti et al., 2021, p. 6).

In this context, it is essential to refer back to the role of health personnel, without whom surrogacy is impossible. Health professionals should provide all necessary information about the risks, whether physical or psychological, to all persons involved in the procedures of artificial insemination and surrogacy (Türk & Terzioğlu, 2014, p. 106). Medical personnel who are involved in the process of artificial insemination can fall into various ethical traps, from the trade of gametes and embryos to the children's trade. Therefore, it is important to note that health personnel must know and realize, take care of, and ensure that all the rights of the embryo and the born child are respected. They must be guided by the basic bioethical principles of harmlessness, beneficence, and, above all, the principles of justice and autonomy. Under no circumstances should they engage in the trade of human reproductive cells, embryos, or women and children, nor is it legally or ethically permissible to exploit individuals, women, and children without informed consent for scientific purposes or the benefit of personal scientific success. As Donum vitae (1987, item 2.7) stated, "*The medical act must be evaluated not only with reference to its technical dimension but also and above all in relation to its goal which is the good of persons and their bodily and psychological health.*"

Conclusion

Surrogacy, as the most delicate (by)product of modern medical achievements, inevitably separates motherhood and parenthood from the traditional concept of mother and family. It is precisely the departure from traditional postulates and the social presumption that the mother of the child is the woman who gave birth to it that the opponents of surrogate motherhood "condemn" because pregnancy is reduced to a mere business transaction devoid of emotional connection and inherent biological connection. However, in an atmosphere of technological progress that provides opportunities to achieve parenthood through surrogacy, it is clear that domicile bans on surrogacy have a minimal effect on dissuading intended parents from resorting to such a practice. This is because the "promised" countries where such practices are legal enable the realization of the desire for biological offspring.

Despite numerous challenges and problems of a (bio) ethical nature related to surrogacy that require an interdisciplinary approach to solving, the law must adapt to technological changes and medical achievements to solve the complex issues arising from the practice of surrogacy. The above certainly requires discussions at the academic level and monitoring of the surrogacy process in a legal and bioethical context in all phases of its realization. We must not lose sight of the fact that although surrogacy, as a blessing of modern medicine, enables the realization of parenthood for some, for others, it can represent a violation of human rights, a negation of human dignity, and child commodification.

References

Ashley, B. M. (2006). *Health care ethics: A Catholic theological analysis*. Georgetown University Press.

Caster, A. (2010). Do not split the baby: How the US could avoid uncertainty and unnecessary litigation and promote equality by emulating the

British surrogacy law regime. *Connecticut Public Interest Law Journal, 10*(2), 139–184.

Coalition Internationale pour l'abolition de la Maternité de Substitution (CIAMS). (2022). *Feminist critical document on the Verona principles*. Retrieved November 21, 2023, from http://abolition-ms.org/en/our-actions/critical-document-on-the-verona-principles/

Frković, A. & Katalinić, S. (2006). Pušenje i alkohol u trudnoći. Pitanje sukoba interesa majka—plod. *Gynecologia et perinatologia: Journal for Gynecology, Perinatology, Reproductive Medicine and Ultrasonic Diagnostics, 15*(3), 166–171.

Frković, A. (2010). *Medicina i bioetika*. Pergamena.

González, N. I. (2020). Legal and ethical issues in cross-border gestational surrogacy. *Fertility and Sterility, 113*(5), 916–919. https://doi.org/10.1016/j.fertnstert.2020.03.003

Harris, L. H. (2000). Rethinking maternal-fetal conflict: Gender and equality in perinatal ethics. *Obstetrics & Gynecology, 96*(5), 786–791. https://doi.org/10.1097/00006250-200011000-00028

Iliadou, M. (2018). Surrogacy and the ECtHR: Reflections on Paradiso and Campanelli v Italy. *Medical Law Review, 27*(1), 144–154. https://doi.org/10.1093/medlaw/fwy002

Inhorn, M. C., & Patrizio, P. (2009). Rethinking reproductive "tourism" as reproductive "exile." *Fertility and Sterility, 92*(3), 904–906. https://doi.org/10.1016/j.fertnstert.2009.01.055

Jadva, V., Murray, C., Lycett, E., MacCallum, F., & Golomboket, S. (2003). Surrogacy: The experiences of surrogate mothers. *Human Reproduction, 18*(10), 2196–2204. https://doi.org/10.1093/humrep/deg397

Macklin, R. (1991). Commentary: Ethical issues in surrogate motherhood. *Women's Health Issues, 1*(3), 148–150.

Margalit, Y. (2014). In defense of surrogacy agreements: A modern contract law perceptive. *William & Mary Journal of Women and the Law, 20*(2), 423–468.

Pande, A. (2010). Commercial surrogacy in India: Manufacturing a perfect mother worker. *Signs, 35*(4), 969–992. https://doi.org/10.1086/651043

Patel, N. H., Jadeja, Y. D., Bhadarka, H. K., Patel, M. N., Patel, N. H., & Sodagar, N. R. (2018). Insight into different aspects of surrogacy practices. *Journal of Human Reproductive Sciences, 11*(3), 212–218. https://doi.org/10.4103/jhrs.JHRS_138_17

Piersanti, V., Consalvo, F., Signore, F., Del Rio, A., & Zaami, S. (2021). Surrogacy and "procreative tourism". What does the future hold from the ethical and legal perspectives? *Medicina, 57*(1), 1–16. https://doi.org/10.3390/medicina57010047

Principles for the protection of the rights of the child born through surrogacy (Verona principles). (2021). International Social Service—General Secretariat.

Retrieved November 21, 2023, from https://www.iss-ssi.org/wp-content/uploads/2023/03/VeronaPrinciples_25February2021-1.pdf

Radan, M., Vuletić, S., Rakošec, Ž., & Šperanda, Ž. (2015). Bioetička kompleksnost problematike zamjenskoga majčinstva. *Diacovensia: teološki prilozi, 23*(1), 35–70.

Roksandić Vidlička, S., Hrstić, D., & Kirin, Z. (2012). Bioethical and legal challenges of surrogate motherhood in the Republic of Croatia. *Jahr, 3*(1), 37–67.

Ryznar, M. (2009). International commercial surrogacy and its parties. *John Marshall Law Review, 43*(4), 1009–1040.

Spivack, C. (2010). The law of surrogate motherhood in the United States. *The American Journal of Comparative Law, 58*(1), 97–114. https://doi.org/10.5131/ajcl.2009.0042

Türk, R., & Terzioğlu, F. (2014). Ethical issues in the surrogate maternity practice. *Caucasian Journal of Science, 1*(1), 98–108.

Večernji.hr. (2023). *Bivša djevojka Davora Šukera nije postala majka u 69. godini: 'Ispunila sam želju pokojnog sina'*. Retrieved November 21, 2023, from https://www.vecernji.hr/sport/zena-koja-je-odbila-sukera-nije-postala-majka-u-69-godini-ispunila-sam-zelju-pokojnog-sina-1669974

CHAPTER 8

Future of the Abortion in Croatia

Ivana Tucak, Josip Berdica, and Laura Seleš

Abstract In Croatia, 2024 was characterized by the unwillingness of the legislature to regulate women's right to abortion with appropriate law. In the Republic of Croatia, abortion is regulated by the Law on Health Measures in Implementation of the Right to Decide on Childbirth, adopted in 1978. The law was passed when the Republic of Croatia was still part of the Socialist Federal Republic of Yugoslavia. Owing to the new constitutional and legal framework after the establishment of the Croatian state, it is clear that the law mentioned above needs to be updated and adapted to the new order. In its Ruling of February 21, 2017, the Constitutional Court ordered the Croatian Parliament to pass a new law on abortion within two years. Since there are no sanctions for the Parliament concerning noncompliance with this deadline,

Present Address:
I. Tucak (✉) · J. Berdica · L. Seleš
Faculty of Law, Josip Juraj Strossmayer University of Osijek, Osijek, Croatia
e-mail: itucak@pravos.hr

J. Berdica
e-mail: jberdica@pravos.hr

L. Seleš
e-mail: lauraseless@gmail.com

I. Pavić et al. (eds.), *Bioethics and Social Ethics in The Modern World*,
https://doi.org/10.1007/978-3-031-86418-6_8

it is not surprising that six years after the decision of the Constitutional Court, a new law has not yet been adopted. This reluctance of the legislature is because it is an issue that deeply divides the nation. The purpose of this paper is to investigate what options are available to the Croatian Parliament when a new law on abortion is adopted. The new law must comply with the Croatian constitutional order and internationally accepted obligations in protecting human rights. Contemporary trends in the regulation of abortion in the member states of the European Union and the United States of America and their possible influence on Croatian legislation will be considered.

Keywords Constitutional Court of Croatia · Abortion · Reproductive rights · European court of human rights · The Croatian Abortion Act

Introduction

Abortion, apart from being a moral, ethical, philosophical, legal, medical, and psychological issue, is also a fundamental sociological issue. The "problematic issue of abortion" certainly did not bypass the Croatian academic community, which at the same time means that abortion is an important issue for society as a whole. Discussions about abortion include moral-ethical, philosophical, theological, sociological, and legal dimensions, which often diverge in societies of "reasonable pluralism," as described by John Rawls, both in progress and even more so in conclusions. This often leads to mutual misunderstanding among members of the academic community and wider society regarding important issues such as the right to life of the fetus on the one hand and the right to freedom (of women to decide) on the other hand. If we add the ideological dimension to this entire discussion, the "problematic issue of abortion" seems almost intractable.

This article aims to present the challenges currently facing the Croatian legislature in its obligation to regulate the issue of abortion. In the Republic of Croatia, abortion is currently regulated by a law that was passed during the socialist era almost half a century ago. In 2017, the Constitutional Court ordered the Croatian Parliament to pass a new law, giving it an instructional deadline of two years, which was not met. The paper is divided in the following way. First, a historical overview of the development of Croatian abortion legislation is provided, followed by statistical analyses of performed abortions, which are good indicators of the effectiveness of abortion legislation, and an analysis of public opinion on abortion. The fourth section of the paper examines the arguments presented in the 2017 Constitutional Court case, which questioned the constitutionality of the current abortion legislation. Finally, the paper analyses current proposals for a new Croatian abortion act.

A Brief History of Abortion in Croatia

In Croatia, a law from a country that no longer exists (Yugoslavia) has still been in force since 1978—Act on Health Measures on the Exercise of the Right to the Freedom of Decision-Making on Giving Birth (hereinafter Abortion Act, Official Gazette of the Socialist Federal Republic of Yugoslavia, hereinafter SFRY, No.18/1978, 31/86, 47/89,and 88/09). It was preceded by a series of legal acts, decrees, and resolutions that sought to regulate the issue of abortion.

In socialist Yugoslavia, abortion for medical reasons was decriminalized for the first time in 1952 to prevent illegal abortions that endangered the lives of many women (Bonfiglioli & Žerić, 2023, p. 149). In 1952, the Decree on the procedure for performing an authorized abortion was passed (Official Gazette of the Federal People's Republic of Yugoslavia, here and after FNRY, No. 4/52). However, a year earlier, the Criminal Code had proscribed abortion and infanticide (Official Gazette of the FNRY, No. 13/51, Articles 138 and 140).

The complete liberalization of abortion up to the 10th week of pregnancy at the request of a pregnant woman occurred in 1969 (Bonfiglioli & Žerić, 2023, pp. 151–152). In the first half of 1969, the General Act on Termination of Pregnancy (Official Gazette of the SFRY, No. 25/69) was adopted together with the Family Planning Resolution (Official Gazette of the SFRY, No. 25/69). The resolution has the following passage:

> *The changed position of women in society changes the structure of the family and family relationships and significantly affects the realization of the motherly function of women. One of the aspects of these changes is a reduced number of births and a reduction in the death rate of newborns. On the other hand, the size of the family today is far more connected with the realization of the desire for a more humane life and is an integral part of people's*

> *efforts to achieve a better standard of living. Deciding on family size is in the common interest of parents and society [...] Family planning as a permanent human and social effort, in its modern application, ensures significant social, health, economic, political and other advantages. (Art. I)*

This resolution emphasized the importance of family and family planning. However, the Resolution does not fail to indicate that induced abortion is still widely used for birth control. Induced abortion is harmful to a woman's health and life-threatening, while at the same time, the spread of contraception is very slow, says Resolution. We should consider that we are talking about Yugoslav society from the end of the 1960s. Specifically, in Britain, for example, it was not until 1974 that contraception was generally available through physicians in the National Health Service (Richards, 2004, 489). Thus, even at Yugoslavia at the time, full attention to "*spreading contraception and encouraging activities in this direction*" proved to be "the most suitable path in further family planning. "*On this path, the education system, as an integral part of health prevention measures to prevent unwanted pregnancies, should have been significant help.*"

As is often the case here, too, the law lags behind social analyses and recommendations. However, this did not prevent the legislature from defending the legal right to abortion against the pressure caused by the then-current encyclical Humanae Vitae (1968) of Pope Paul VI, which called for a "*complete ban on abortion.*"[1] The Yugoslav official policy at the time rejected such a position of the Catholic Church in general, considering it an attempt to interfere with the affairs of the state, the position of the family, and the self-governing rights of citizens and institutions, which is not following the principle of complete separation of the church from the state as declared in the SFRY Constitution. The content of the Resolution from 1969 received its highest legal expression in the

[1] We can read: "*[T]he direct interruption of the generative process already begun and, above all, all direct abortion, even for therapeutic reasons, are to be absolutely excluded as lawful means of regulating the number of children*" *(*Paul VI 1968, *No. 14). In the corresponding footnote (14) we can look at the history of this position of the Magisterium of the Church (see also:* Mapasure & Rakoczy 2016*). In addition, today's Pope* Francis (2018) *says in one of his catechesis:* "*A contradictory approach even permits the termination of human life in the maternal womb, in the name of safeguarding other rights. However, how can an action that ends an innocent and defenseless life in its blossoming stage be therapeutic, civilized or simply human? I ask you: is it right to 'do away with' a human life in order to solve a problem? Is it right to hire a hitman in order to solve a problem?*".

1974 SFRY Constitution (Official Gazette of the SFRY, No. 9/74, Art. 191):

> *It is a person's right to freely decide on having a child.*
> *This right may be restricted only for reasons of health.*[2]

According to prominent authors, this request for free parenthood, i.e., for free family planning, contains at least two essential components: every individual's fundamental human right to decide "*on the number and timing of having children*" (Alinčić, 1976, p. 363). Second, it is an integral and inseparable element of ensuring social conditions and free family planning. Free parenting does not exist and cannot exist if it is conditioned by coercion, misery, poverty, social, and other inequalities and if circumstances are not created that would allow everyone to find alternative solutions. Therefore, free family planning involves two essential elements: *individual* and *social* (Mladenović, 1987, p. 262). We can understand the adopted legal solution in such a conceptual model.

The content of Article 191 of the SFRY Constitution was also included in the constitutions of all republics and provinces that make up Yugoslavia (Perinčić, 1990; Hlača, 2009, p. 143). It is identical to Article 272 of the Socialist Republic of Croatia Constitution (Official Gazette, No. 9/74). Free family planning was enacted in 1978 in the then-Socialist Republic of Croatia. Although the title Act on Health Measures in Family Planning was initially proposed, the legislator decided to make its essential purpose unequivocal; hence, the title of the new act explicitly contains the name of the right it protects—the right to the freedom of decision-making on giving birth.

Notably, the Abortion Act of 1978 was not adopted according to the "*liberal anthropological matrix of the Western world*," nor was it the result

[2] For the translation of this article of the constitution, we consulted Alinčić, 1976, p. 363.
The proposal that preceded this constitutional provision was: "*To achieve the freedom of a woman's personality, a woman's right to terminate a pregnancy is ensured. This right can be limited only in the interest of protecting a woman's life and health*" (Petrić 1981, 334). However, this proposal was not adopted because the "*right to abortion*" *as a constitutional category would have a disincentive effect on society's obligation to develop social preconditions—education, information, social and health care—which, in the spirit of the Resolution from 1969, would helped families to independently decide on the number of children and the time of their birth*".

of political compromises of key actors at the time since there was no multi-party democracy in socialist Yugoslavia (Popović, 2018, 132). Act results from "*the promotion of socialist goals, then paradigmatic in these areas, which sought to expand the availability of women as much as possible for her fuller inclusion in the context of 'self-governing' economic production*" (Popović, 2018, p. 132).

Abortion Statistics

The effectiveness of an abortion law can be assessed by its impact on the number of abortions performed (Hlača, 2009). According to the Abortion Act (Art. 15, Para. 2), in the Republic of Croatia, a woman can legally have an abortion without the commission's permission if it is a pregnancy that lasts up to ten weeks. In accordance with the same Act, abortion can be performed only in authorized inpatient healthcare facilities or hospitals.

What can we say about abortions in 2023, according to official data? According to the data of the *Croatian Institute of Public Health (Hrvatski zavod za javno zdravstvo)*, a total of 7,807 abortions were reported in Croatian hospitals in 2023, which represents a decrease of 7.4% compared with the previous year (Table 8.1). Among the total number of spontaneous abortions, 38.6% (3 015) were legally induced, while spontaneous abortions made up 18% of the total number. Among the women in whom a legally induced abortion was registered, the largest share were those aged 30–39 years–44.3%, followed by those aged 20–29 years, in which 38.5% of all legally induced abortions were registered. Sixty-one legally induced abortions in minors (up to the age of 18) were recorded, accounting for 2.0% of the total number of legally induced abortions. Another 144 legally induced abortions were recorded among young girls between the ages of 18 and 20, which together make up 6.8% of the total number of legally induced abortions recorded.

Table 8.1 Number of legally induced abortions in Croatia (1990–2023)

Years	*1990*	*1995*	*2000*	*2010*	*2020*	*2023*
Number of legally induced abortions	38 644	14 282	13 870	10 150	7 485	7 807

Source Croatian Institute of Public Health, 2024

These numbers, especially in recent years, require a certain amount of caution since some authors note that only one-third of the actual number of abortions are reported in Croatia. In contrast, others will talk about fifty percent of the reported abortions. Unregistered abortions most often occur not only in private surgeries but also in nonprofessional institutions. This step is taken mainly by women who want to remain anonymous or who want to have an abortion outside the legal frame. There is also a considerable number of medical personnel who see abortions as a personal source of profit. Specifically, an abortion that is not the result of some pathology is a medical procedure that the woman pays for herself. In simpler terms, in Croatia, there has not been a significant reduction in the number of abortions within thirty years. However, many of these methods have been performed illegally.

Some authors (Šijanović & Kovačević, 2011, 171–172) consider this visible drop in the number of abortions since the 1990s as expected and cite the following explanations: a change in the social attitude toward abortion (an immoral act that should be prohibited); and better education, information, and availability of contraception. Despite the apparent progress, Croatia, as a member of the EU, is ranked quite low on the so-called "Contraception Policy Atlas Europe 2024."[3] The atlas contains a complete interactive map showing the availability of contraception in European countries and is published by the European Parliamentary Forum on Sexual and Reproductive Rights. We are talking about the right to abortion, while at the same time, we do not have a developed awareness of reproductive health and safe family planning. We still understand abortion as a form of family planning. Although the number of registered abortions is decreasing, the number of those who do not want to ban it is not. According to the HRejting survey (2022), only 15% of respondents

[3] This is particularly problematic if we take into account the clear position of the Parliamentary Assembly of the Council of Europe and Resolution 2331 (2020) about empowering women ("Promoting Access to Contraception in Europe") where we read: "*The Assembly calls on [...] to ensure that all modern methods of contraception, including emergency contraception without prescription, are made available to the public, including in rural and remote areas and to marginalized and vulnerable groups; [...] ensure the affordability of contraceptive methods by including them in national health insurance schemes [...] develop specific coverage or subsidization schemes for young, low-income and vulnerable groups, [...] review such schemes regularly to ensure their effectiveness; [...] provide affordable, confidential and nonjudgmental individual counseling to those seeking contraception*" (https://assembly.coe.int/nw/xml/XRef/Xref-XML2HTML-en.asp?fileid=28675&lang=en).

in Croatia believe that abortion should be completely banned. For even 73% of the population, the ban is unacceptable. Approximately 11% of those surveyed are neutral; they neither agree nor disagree. According to the same survey, belonging to different ideologies (in Croatia, these are left-wing or right-wing conservative voters) has no significant influence on the distribution of votes on the issue of abortion in Croatia.[4]

Debate on Values

What we may see in our short historical review is, among other things, the focus of legislators on women and their health. For a long time, the abortion debate was conducted about its social consequences, the problem of backstreet abortions, and other negative consequences for women and their loved ones (Holm, 2023, p. 3). These discussions were related to the protection of public health. Nevertheless, since the 1960s, at the beginning of the feminist movement, the issue of abortion has increasingly been viewed in the context of sexual and reproductive freedoms and human rights (Holm, 2023, p. 3).

Importantly, today's discussions are focused on the woman and her health and the fetus. The development of modern medicine has also contributed to this in part. The development of neonatology and intensive care for newborns has advanced the "age of viability." Abortion of a "potentially viable fetus" is generally considered inadmissible, with the usual exceptions when the continuation of pregnancy poses a serious health risk to the mother or when a serious health threat to the fetus has been discovered (Holm, 2023, p. 341). In this context, ideas have developed about how the fetus's lack of legal subject status can be compensated for by recognizing its patient status (Lyerly et al., 2008). In sum, today's debates concern the value of the life of a woman/mother and a fetus/child. It concerns values for which we have different attitudes (Table 4.1.). Some authors will view liberalism as the source of modern debates on abortion. "Children are increasingly viewed as a limitation upon individual freedom, which contributes to liberalism's commitment to abortion on demand," says Deneen (2018, p. 39). The discussion about contemporary values and abortion necessarily leads us to the question of freedom, which here, too, proves to be the concept that is the

[4] https://vijesti.hrt.hr/hrvatska/hrejting-za-73-ispitanika-zabrana-pobacaja-neprihvatljiva-7502302.

Table 8.2 Three Views of Abortion

Status of unborn	*Fully human*	*Potentially human*	*Subhuman*
Abortion	Never	Sometimes	Anytime
Basis	Sanctity of life	Emergence of life	Quality of life
Mother's rights	Life over privacy	Combination of rights	Privacy over the right to life

Source Geisler, 2005, 133

most difficult to clarify among all political concepts (Barry, 1995, 204) (Table 8.2).

Despite different understandings and representations of different values, these questions and dilemmas are present almost daily in contemporary Croatian society. The Act, which has been in force in Croatia since 1978, is certainly not such high quality that its "lifetime" would last almost half a century. The reasons are much more complex than the mere passing of a new law because the values that dominate Croatian society are at stake. These values, especially in regard to abortion, are still dominantly conditioned by religion, more precisely, by Catholic (moral and social) teachings. In addition to the fact that almost eighty percent (78,97%) of citizens still declare themselves Catholic in 2022 Croatia, the presence and importance of the Catholic Church in the matter of abortion are very important. According to the "European Values Study 2017" results, religion and religious practices "*significantly influence the value attitude toward abortion.*" Respondents who attend religious ceremonies tend to disapprove of abortion more often than do those who attend them less often or not at all. Thus, 63% of the respondents who attended religious ceremonies at least once a week or more never approved of abortion. Among those who attend religious services once a year or less often, almost 24% will always approve of abortion (Matulić & Balabanić, 2019, 392).

Constitutional Court and Abortion

The Croatian Abortion Act is a scarce example of a Croatian law that has reached its 47th birthday.[5] The act was based on the Constitution of the SFRY from 1974, which guaranteed the freedom of the individual to decide on the birth of children. The same constitutional principle could also be found in the 1974 Constitution of the Socialist Republic of Croatia (Hlača, 2009, p. 143). The new Croatian Constitution does not contain a similar provision. However, the constitutions of some countries that, like Croatia, were created by the breakup of Yugoslavia do so. Today, such an express provision on the freedom of an individual to decide on the birth of children can be found in the constitutions of four such states: Kosovo, North Macedonia, Serbia, and Slovenia (Śledzińska-Simon, 2023, p. 440).[6] Authors who do not know the constitutional law of the former Yugoslavia mistakenly attributed the sources of these constitutional provisions to the intervention of contemporary international experts in their adoption (Śledzińska-Simon, 2023, pp. 400–401).

The greatest challenge to the continued validity of the controversial Abortion Act thus far has been the constitutional court proceedings to review its conformity with the new Croatian Constitution. However, it is important to note that it took the Constitutional Court twenty-six years to decide on this issue, as the first proposals for the institution of constitutional proceedings to review its conformity with the Constitution began arriving immediately after the adoption of the new Constitution (Popović, 2018, p. 133).

On February 21, 2017, the Constitutional Court issued a ruling rejecting the proposals for the assessment of constitutionality. Nevertheless, it ordered the Croatian Parliament to update abortion legislation to

[5] Croatian family law expert Nenad Hlača expressed his surprise in his 2009 article that the Abortion Act had reached its thirtieth birthday. In addition to abortion, the Abortion Act also regulates the issues of contraception and sterilization (Articles 5–14) and "medical assistance in cases of reduced fertility", i.e., artificial insemination of a woman with her husband's or another man's sperm, which in the 1970s was the only possible form of assistance for infertile persons (Articles 29–34, 42 paragraph 1, point 3). The provisions of the Abortion Act relating to artificial insemination ceased to apply with the entry into force of the special Act—Zakona o medicinskoj oplodnji (Official Gazette No. 88/09). See Hlača, 2009, p. 143.

[6] See Kos. Const. art. 26(2) (2008, as amended); North Macedonia N. Maced. Const. art. 41 (1991, as amended); Serb. Const. art. 63, § 1 (2006); Slovn. Const. art. 63 (1991, as amended).

adopt a new law within two years.[7] However, in 2025, there is still no indication that the parliament or the current ruling coalition is working seriously on fulfilling this obligation.

It must be emphasized at the beginning of this section that Croatia's need to regulate abortion legislation is not the result of some new "conservative revolution" but rather decades of "sweeping under the carpet" of this critical issue related to the protection of human rights, which must be regulated following the new Croatian Constitution, international treaties concerning the protection of human rights, and the practice of the bodies implementing these treaties (Popović, 2018, pp. 133, 135).

In its ruling, the Constitutional Court was guided primarily by the teleological or objective method of interpreting the constitutional text (Popović, 2018, p. 135). This method implies "the correct interpretation of the goal, or rather the purpose for which the legal norm is intended in today's current dynamic-evolutionary context of its application" (Popović, 2018, p. 135, n. 12). That is, it was not guided primarily by the historical method that explores *travaux préparatoires* and the procedure by which the act was adopted to determine the true intentions of its creators (Popović, 2018, p. 135, n. 12).

The issue of abortion now regularly appears before the supreme or constitutional courts of various countries. In the European context, we already have a rich case law of the European Court of Human Rights and the somewhat more modest European Court of Justice (See Tucak & Blagojević, 2020). This is the result of the fact that "*the abortion question is no longer framed as a question of how to solve social problems and thereby a primarily political problem (on which it is possible to reach a compromise) but as a question of how to implement overriding human rights*" (Holm, 2023).

Sexual and reproductive rights (SRRs) are currently protected by the most important international and European human rights documents, of which the International Covenant on Civil and Political Rights (1966), the International Covenant on Economic, Social and Cultural Rights (1966), the UN Convention on the Elimination of All Forms of Discrimination against Women (1979), the European Convention on Human

[7] Ruling of the Constitutional Court of the Republic of Croatia No. U-I-60/1991 et al. of February 21, 2017 and Separate Opinion, available at: https://narodne-novine.nn.hr/clanci/sluzbeni/2017_03_25_564.html. See also analysis Popović, 2018, pp. 133, 135.

Rights (1950), and the Charter of Fundamental Rights of the European Union (2000) deserve to be mentioned (Resolution 2020/2215 (INI), p. C, Holm, 2023). Access to abortion, as emphasized by Soren Holm, "*has become a primarily legal issue involving a conflict of rights in which the 'winner takes all'*" (Holm, 2023, p. 342).

Court interpretations establishing the right to abortion appear to be a response to the constitutional silence on this issue (Śledzińska-Simon, 2023, p. 402). In these cases, judges find the existence of this right with the help of other constitutional guarantees: "*liberty, equality, dignity, or, more recently, the protection of health*" (Śledzińska-Simon, 2023, pp. 399, 402).

In this paper, we first investigate the arguments presented in numerous "*proposals for the institution of proceedings to review the conformity*" of the Abortion Act with the Constitution. Then, the arguments of the Constitutional Court itself are presented, more precisely, the majority opinion.

The Abortion Act was questioned by numerous proponents, for example, the Croatian Movement for Life and Family, the Croatian Catholic Association "WE," the Association "In the Name of the Family" and private citizens. The Croatian Movement for Life and Family noted that the current law that allows forcible termination of pregnancy does not follow the Constitution because the Constitution guarantees that every human being has the right to life (Constitution of the Republic of Croatia, 1990: Article 21), emphasizing that an almost identical provision we find in the UN Universal Declaration on Human Rights: "*Everyone has the right to life, liberty and security of person*" (Universal Declaration on Human Rights, 1948: Article 3). The current law is based on nonscientific facts that a newly conceived child is not a new person of its own and that abortion is contrary to medical ethics (Ruling of the Constitutional Court of the Republic of Croatia and the Dissenting Opinion, 2017, paragraph 4.1). The proponent, the Croatian Movement for Life and Family, noted that the Abortion Act is also unacceptable from a moral point of view because one human living being decides on the life and death of another human living being. Under the aforementioned approach, referring to the Constitution according to which everyone is equal before the law (Constitution of the Republic of Croatia, 1990: Article 14), the proponents consider abortion a negation of rights, not a right (Ruling of the Constitutional Court of the Republic of Croatia and the Dissenting Opinion, 2017, paragraphs 4.1 and 4.1.1).

The proponent, the Association "In the Name of the Family," considered the Abortion Act unconstitutional also concerning Article 16 of the Constitution, according to which freedoms and rights may be restricted only to protect the freedoms and rights of other people, the legal order, public morals, and health, and that such restriction must be proportionate "to the nature of the need" (Constitution of the Republic of Croatia, 1990: Article 14). Additionally, the Abortion Act, according to this proponent, is incompatible with articles 23, 59, and 65 of the Constitution (prohibition of abuse, right to health care, duty to protect children, and infirm persons). The provision of the Abortion Act, which stipulates that termination of pregnancy is carried out at the request of the pregnant woman (Abortion Act: Article 18), was highlighted as controversial, explaining this argument by the intention of the Croatian lawmaker to put both parents in an equal position regarding their child (Ruling of the Constitutional Court of the Republic of Croatia and the Dissenting Opinion, 2017, paragraph 4.6).

The proponent, a private citizen, Saša Čajić, noted the incompatibility of the contested Act with the UN Convention on the Rights of the Child (1989: Article 6), which states that contracting states ensure the survival and development of the child to the greatest extent possible and recognize every child's innate right to life. Nevertheless, here, the following question must be asked: what exactly does the natural right mean? Fundamental human rights are created at birth, and it is utterly irrelevant whether the child is only one minute old (Vilić & Tucak, 2023, p. 25). While still in the mother's womb, the fetus has no human rights, but after birth, it becomes the bearer only a few minutes later. The UN Convention on the Rights of the Child, which is the product of a compromise between states, defines a child as every human being under 18, unless under the law of a particular country, the majority is attained earlier (Article 1). The Convention nowhere mentions the legal personality of the unborn child and thus does not guarantee them the rights contained therein (Hlača, 2009, p. 145).

In the proceedings before the Constitutional Court, the opinions of experts from various scientific fields were obtained: law, theology, medicine, and medical ethics. The Chair of Family Law of the Faculty of Law in Zagreb noted that the fundamental problem of the Contested Abortion Act is the question of whether an unborn child has an a priori right to life, whether the attribute of a human being should be assigned to an unborn child or whether his mother has the freedom to self-determination. They emphasize that a human embryo cannot be denied the status of a human being (Ruling of the Constitutional Court of the Republic of Croatia and the Dissenting Opinion, 2017, paragraph 5.2). The latter is an indisputable scientific fact from the point of view of biomedical sciences because every human being begins to exist at the moment of conception and, as such, is a new, unique, and unrepeatable biological organism (Hrabar, 2015, p. 820).

What is certain is that fetuses, that is, the unborn, are human beings, but whether they are also persons is a key question in abortion regulations. As mentioned, birth is still the key moment in all modern legal systems for recognizing legal subjectivity (Vilić & Tucak, 2023, p. 19). The key question is whether the fetus should be recognized as a person with the right to life (Vilić & Tucak, 2023, p. 22). Professors from the Chair of Family Law at the Faculty of Law in Zagreb stated that from a biomedical point of view, it is an indisputable fact that life begins at conception and that if the framers of the Constitution thought differently, they should have guaranteed the right to life only to an already born person (Ruling of the Constitutional Court of the Republic Hrvatska and the Dissenting Opinion, 2017, paragraph 5.2). They agreed that the legislator should balance the conflicting interests of the right to life guaranteed in the Constitution and the woman's right to free choice (Ruling of the Constitutional Court of the Republic of Croatia and the Dissenting Opinion, 2017, paragraph 5.2).

Theology experts gave an opinion in their exposure to the Constitutional Court that the Abortion Act should be entirely repealed with the argument that human life begins at conception (Ruling of the Constitutional Court of the Republic of Croatia and the Dissenting Opinion, 2017, paragraph 5.5). Experts in medical ethics refer to the medical oath in its original version from 1948: "*I will respect human life from the very beginning*" (The World Medical Association Declaration of Geneva, Physician's Oath, 1948; Ruling of the Constitutional Court of the Republic of Croatia and the Dissenting Opinion, 2017, paragraph 5.6).

However, it is important to note that the Pledge of the World Medical Association has changed several times over the past decades. Modern pledge versions do not prohibit abortion (Hajar, 2017). The medical oaths taken by today's doctors no longer mention the doctor's obligation from the original oath to "do no harm" or not to give patients "lethal medicine" (Hajar, 2017).

Some authors suggest that instead of the term "unborn child," it is better to use the term "patient" (Lyerly et al., 2008). It is believed that this avoids pitfalls when the term fetus or unborn child is used by emphasizing "*the beneficence—rather than the rights-based nature of obligations toward early life*" (Lyerly et al., 2008). According to the professor from the Zagreb Chair of Family Law, Dubravka Hrabar, an unborn child is considered a patient, and the fetus is a patient like any born person; the only thing is that it is in utero (Hrabar, 2015, p. 827). However, what exactly does it mean to be a patient from a medical point of view, what criteria does someone have to fulfill to be considered a patient, and is it possible to apply the same criteria to the fetus as an entity separate from the pregnant woman? Being a patient means benefiting when the members of the medical profession are applying their clinical (medical) skills to you. The conditions for realizing such a benefit are that it is a human being involved and that the medical methods used are expected to be helpful for a particular entity (person, patient) (Kurjak et al., 2009, pp. 344–345). As the fetus is undoubtedly a human being and medical intervention can be used to help him, according to the aforementioned methods, we can talk about the fetus as a patient.

This is a controversial topic. For example, the authors in the paper "A critique of the 'fetus as patient'" state that the term patient is unpragmatic when talking about the fetus since it is not an independent physically separate entity, and they posed the question should the status of a patient be recognized only when it is clear that the pregnancy is going to continue to the very end (Lyerly et al., 2008, pp. 1–2). Although it is interesting, it is also essential to think about the above propositions, especially when we consider all the aspects necessary for the fair regulation of abortion; we would not go into this part further, mainly because it is an extensive topic and we do not have enough space to go into the details.

In its assessment, the Constitutional Court noted that abortion is a profoundly moral, ethical, medical, philosophical, legal, and religious issue on which experts from different fields do not have a unanimous opinion and that the central question of the moral permissibility or inadmissibility of abortion has still not been answered (Ruling of the Constitutional Court of the Republic of Croatia and the Dissenting Opinion, 2017, paragraph 21).

The Constitutional Court noted that international law does not answer the question of when life begins and that the protection of human life in international documents is principled, so individual states are primarily responsible for interpreting life's beginning (Ruling of the Constitutional Court of the Republic of Croatia and the Dissenting Opinion, 2017, paragraph 26).

Importantly, the Croatian Constitutional Court, when making its decision, also took into account the case law of other European constitutional courts. It gave in its Ruling, among others, an overview of the case law of the constitutional courts in Hungary, Slovakia, Germany, and Spain. Our Constitutional Court ultimately took a position on abortion, which is very similar to the position of the Hungarian Constitutional Court. The Constitutional Court of Hungary, in its decision number 64/1991 of December 17, 1991, said the following:

> *The question of life protection and the regulation of abortion is exclusively in the domain of the Parliament. Whether a fetus is a subject of law or not is a political decision (...). The complete abortion ban is not constitutional because it denies the pregnant woman's right to self-determination, as would the rules that would give priority exclusively to that right. The state is obliged to protect life from the very beginning, and this duty implies that the state cannot legally allow unfounded terminations of pregnancy* (Ruling of the Constitutional Court of the Republic of Croatia and the Dissenting Opinion, 2017, paragraph 18).

On the basis of everything presented, the Croatian Constitutional Court concluded that although the right to decide a child's birth freely is no longer expressly guaranteed by the new Constitution, it is protected by the right to privacy from Article 35. The Constitutional Court concluded that the right to privacy guarantees a woman "*the right to decide on conceiving a child and on the course of her pregnancy*" (Ruling of the Constitutional Court of the Republic of Croatia and the Dissenting Opinion, 2017, paragraph 44.1). By becoming pregnant, a woman does not lose her right to free, autonomous decision-making, i.e., self-determination (Ruling of the Constitutional Court of the Republic of Croatia and the Dissenting Opinion, 2017, paragraph 44.1). However, like most constitutional rights, the right to privacy is not absolute, so it can be legally limited to achieving one or more legitimate social goals

(Ruling of the Constitutional Court of the Republic of Croatia and the Dissenting Opinion, 2017, paragraph 44.1.).

Regarding the fetus's right to life, the Constitutional Court concludes:

> *In this respect, the Constitutional Court finds that an unborn being, as a value protected by the Constitution, enjoys constitutional protection under Article 21 of the Constitution only to the extent that such protection does not conflict with the woman's right to privacy. The right to life of an unborn being within this meaning is not protected to have an advantage over or greater protection than a woman's right to privacy. Within this meaning, the legislator enjoys the freedom of discretion in striking a fair balance between a woman's right to free decision-making and privacy, on the one hand, and the public interest in ensuring the protection of an unborn being, on the other* (Ruling of the Constitutional Court of the Republic of Croatia and the Dissenting Opinion, 2017, paragraph 45).

Importantly, in legal and philosophical discussions, the observation of the normative status of abortion as a private matter is controversial (Popović, 2018, p. 152). Thus, Popović points to the social significance of the issue of abortion since the "fair balance" of the opposing rights of the woman and the fetus determines who can be considered a member of the social community and a holder of rights (Popović, 2018, p. 152). The question of whether the right to privacy can be adequately used as a guarantee of protection of a woman's free decision-making about abortion has been the subject of numerous debates among constitutional law experts in the USA. This approach has its opponents not only among opponents of the right to abortion but also among advocates of this right of a woman to free choice (for more details, see Tucak & Blagojević, 2023). The Constitutional Court ultimately concluded that it is necessary to modernize the current law on abortion since it is indisputable that some legal institutes and concepts of this law are based on categories that do not exist today (Ruling of the Constitutional Court of the Republic of Croatia and the Dissenting Opinion, 2017, paragraph 35). For these reasons, the Court gave an instruction deadline to the legislature to pass a new act that would solve that problem but would also contain "*educational and preventive measures (…) so that termination of pregnancy is an exception*" (Ruling of the Constitutional Court of the Republic of Croatia and the Dissenting Opinion, 2017, paragraph 50).

What is Next?

Although the government has not yet presented its proposal for a new abortion law, some of the opposition parties have become active in this area. On September 30, 2020, the oppositional Social Democratic Party submitted to the members of the Croatian Parliament, the Draft Law on the Medical Termination of Pregnancy (Prijedlog zakona or medicinskom postupku prekida trudnoće, hereinafter the Draft Law),[8] requesting its enactment under urgent procedures, and in July 2022, together with other left-wing and liberal parties, they tried in the Croatian Parliament to win support for the inclusion of the right to abortion in the Constitution.[9] Both initiatives by Croatian opposition parties were unsuccessful. Thus, Croatia failed to overtake France in its effort to become the first country in the world to explicitly guarantee a woman's right to abortion as a human right.[10]

The Draft Law entered the procedure in 2020 as a proposal that must be decided upon in an urgent procedure. Two years later, it was put on the agenda of the Croatian Parliament. However, since most parliamentarians did not vote for it, the proposal was not accepted (10th convocation of the Croatian Parliament, 11th session of the Croatian Parliament, Voting results, 2020). The draft law contained several provisions causing controversy among the Croatian public, so it is helpful to analyze them. Specifically, Article 2 of the Draft Law states that a minor who has reached the age of sixteen can submit a request for termination of pregnancy without the consent of her parents or other legal representative (The Draft Law, Article 2). As a novelty, the draft law introduces a concept that is not recognized by the current abortion act, which is the

[8] https://sabor.hr/sites/default/files/uploads/sabor/2020-10-01/100508/PZ_35.pdf. A similar legal proposal to regulate abortion was also drafted by the nongovernmental organization Centre for Education, Counseling and Research (CESI). See: https://stari.cesi.hr/attach/_z/zakon_o_sprjecavanju_i_prekidu_trudnoce_25_2_.pdf.

[9] https://www.tportal.hr/vijesti/clanak/foto-video-lijeva-oporba-I-dio-vladajucih-skupili-potpise-za-raspravu-O-pobacaju-U-ustavu-foto-20220708.

[10] In 2024, France became the first country in the world to explicitly guarantee a woman's right to abortion in the Constitution. See Bottini, Eleonora, Bouaziz, Margaux; Hennette-Vauchez, Stéphanie: Enshrining Abortion Rights in the French Constitution: A Global Statement with Little Domestic Substance?, VerfBlog, 2024/3/09, https://verfassungsblog.de/enshrining-abortion-rights-in-the-french-constitution/, https://doi.org/10.59704/cb4c6f83ad0b5367.

medical termination of pregnancy; according to the draft law, it may be performed up to the end of the ninth week of gestation.

This solution follows the WHO Abortion Care Guidelines on safe abortions from 2022. The Guideline in Best Practice Statement 49 indicates that today, there is no single approach to providing abortion services, but plurality approaches coexist (Erdman, 2023, p. 1122). Regulating the use of mifepristone and misoprostol, a combination of two medicines with which medical termination of pregnancy is performed today, is therefore crucial (Erdman, 2023, p. 1122).

In contrast to the medical termination of pregnancy, the usual surgical termination of pregnancy specifies twelve weeks of gestational age as the deadline by which an abortion can be performed (The Draft Law, Article 2). The current law allows abortion to be performed up to ten weeks after the date of conception (Abortion Act, article 15). Interestingly, the number of weeks as part of the terminology in the Draft Law increased from ten to twelve weeks. However, it should be noted that twelve weeks of gestation is equal to ten weeks from conception (Vuletić et al., 2020, p. 222), which is therefore the same standardized period as in the current Act. According to the current abortion act, after the expiration of ten weeks from the day of conception, the first instance commission, or the second instance commission, can approve the termination of pregnancy at the request of the pregnant woman only in exceptional cases prescribed by law (Abortion Act, Articles 22–24). This confirms the time limitation of the right to abortion as a kind of protection of the fetus by the state. Over time, a woman's right to abortion weakens (Rittossa, 2005, p. 975).

The opposition's draft law contained new solutions for the Croatian abortion regime, such as the one in which the costs of termination of pregnancy are borne entirely by the Croatian Health Insurance Fund (Draft Law, Article 28). In the current Abortion Act, this cost is borne by the woman herself (Abortion Act, Article 41). In contrast to the current Abortion Act, the Draft Law introduces conscience objection as a right that physicians can exercise and refer to but under the condition that the responsible persons in health care institutions are obliged to ensure that physicians are willing to perform a procedure (Draft Law, Article 27). Although this document recognizes the conscience objection, in contrast to the current Abortion Act, it aims to introduce a legal provision according to which every hospital (that is, a health institution licensed to perform abortion procedures) would be, at the same time, obligated to provide abortion services.

The conscience objection is in the Croatian Constitution (Constitution of the Republic of Croatia, Article 40). In Croatia, a law already regulates the conscience of health care providers, namely, the Law on Medical Practice. According to its provisions, a doctor may, for reasons of their ethical, religious, or moral beliefs, appeal to their conscience and refuse to carry out diagnostics, rehabilitation, and treatment of a patient if it does not conflict with the rules of the profession and does not cause a threat to the patient's health. They are obliged to inform their superiors about this decision (Law on Medical Practice, Official Gazette, No. 121/03, 117/087, Article 20).

As mentioned at the beginning of this section, the draft law proposed by the opposition parties was not adopted by the Parliament, so we are still waiting for the fulfillment of Parliament's obligations according to the ruling of the Constitutional Court. Here, it is crucial to note that the Constitutional Court, in the decision mentioned above, noted that the legislator has quite broad powers when regulating abortion:

> *The legislator, to enable the woman to determine freely regarding pregnancy and maternity, may set an appropriate deliberation period before a decision on termination or continuation of pregnancy is made, during which she would receive all information on pregnancy and services available (for example, counseling centers and health protection during pregnancy and birth, the working rights of pregnant women and mothers, the availability of nurseries, centers that provide contraception and information on safe sex, and centers that provide counseling before and after pregnancy). It is up to the legislature to determine how the new act will regulate the question of costs resulting from the termination of pregnancy (whether and in which cases the woman will bear them or whether they will be settled at the expense of the state budget), the question of conscientious objection of doctors who do not want to perform terminations of pregnancy, etc.* (Ruling of the Constitutional Court of the Republic of Croatia and the Dissenting Opinion, 2017, paragraph, paragraph 50).

Most of what is stated in this paragraph is directly contrary to the 2022 WHO Abortion Guideline and the recent resolutions of the European Parliament.[11] In particular, we are considering the legal introduction of "an appropriate deliberation period before a decision on termination

[11] World Health Organization, Abortion Care Guideline, 2022 available at https://www.who.int/publications/i/item/9789240039483.

or continuation of pregnancy," allowing the conscientious objection of doctors and the independent bearing of the costs of abortion by the woman to whom this service is provided. The listed solutions are thus directly contrary to the European Parliament resolution of June 24, 2021, on sexual and reproductive health and rights in the EU, in the frame of women's health (2020/2215 (INI)).[12] Although the European Union has no direct competence in sexual and reproductive health and rights (SRHR), cooperation between states takes place through the "open method of coordination," which means that the European Union "invites, encourages and supports Member States in advancing SRHR" (Resolution 2020/2215 (INI), p. D).

The European Parliament adopted this resolution on June 24, 2021, on the basis of a report by the Croatian MEP Predrag Fred Matić, which rapporteured for the European Parliament's Committee on Women's Rights and Gender Equality (FEMM). The resolution is characterized by a broad understanding of sexual and reproductive health and rights and, unlike relevant sources of international law, explicitly defines abortion as a human right.[13]

Almost all the propositions that our Constitutional Court left to the discretion of the legislature are defined by the Resolution as "legal, quasilegal and informal barriers" in accessing abortion services:

> *(…) whereas even when abortion is legally available, there is often a range of legal, quasilegal and informal barriers to accessing it, including limited periods and the grounds on which to access abortion; medically unwarranted waiting periods; a lack of trained and willing health care professionals; and the denial of medical care on the basis of personal beliefs, biased and mandatory counseling, deliberate misinformation or third-party authorization, medically unnecessary tests, distress requirements, the costs involved and the lack of their reimbursement* (Resolution 2020/2215 (INI), p. U.; see also Rupčić, 2023, p. 171).

[12] https://www.europarl.europa.eu/doceo/document/TA-9-2021-0314_EN.html.

[13] Resolutions defines sexual and reproductive health (SRH) as "*a state of physical, emotional, mental and social well-being in relation to all aspects of sexuality and reproduction, not merely the absence of dysfunction, infirmity or mortality, and whereas all individuals have a right to make decisions governing their bodies (…), free from discrimination, coercion and violence, and to access SRH services that support that right and take a positive approach to sexuality and reproduction, as sexuality is an integral part of human existence*" Resolution 2020/2215(INI), p. A; Rupčić, 2023, pp. 155–156, 168.

In Croatia, the issue of conscientious objection is particularly sensitive. Importantly, the Resolution defines conscientious objection as the denial of medical care on the basis of personal beliefs (Rupčić, 2023, pp. 160, 171, Resolution 2020/2215 (INI), p. I). The "so-called conscience clause" is accused of endangering women's lives and rights (Resolution 2020/2215 (INI), p. 37; Rupčić, 2023, p. 173).

In April 2024, the European Parliament adopted a resolution on including the right to abortion in the EU Charter of Fundamental Rights, which "*urges all Member State governments to guarantee access to safe, legal and free abortion care*" (Resolution 2024/2655(RSP) p. 7).[14] This resolution explicitly cites the Republic of Croatia as an example of a country where women are denied the opportunity to have an abortion owing to the (mis)use of conscientious objection by doctors:

> *whereas in countries such as Italy, Slovakia and Romania, a large majority of doctors declare themselves conscientious objectors, making the de facto access to abortion care extremely challenging in some regions, whereas access to timely and appropriate abortion care is being denied in other Member States as a result of practical obstacles, such as in Croatia* (Resolution 2024/2655(RSP) p. T)

According to the text of this Resolution, in Croatia, access to abortion is not "extremely challenging" owing to the appeal of the conscience of doctors, as in countries such as Italy, Slovenia, and Romania, women are "denied" access to abortion.

As we have already indicated, all the measures mentioned by the Constitutional Court have been labeled undesirable by the World Health Organization. In its latest Abortion Guidelines, it emphasizes opposition to "mandatory waiting periods for abortion" (Recommendation 6 Erdman, 2023, p. 1121). Recommendation 22 identifies conscientious objection as a "barrier" to abortion care. In certain circumstances, the idea of conscientious objection is considered indefensible:

> *It is critical that States ensure compliance with regulations and design/organize health systems to ensure access to and continuity of quality abortion*

[14] European Parliament resolution of 11 April 2024 on including the right to abortion in the EU Fundamental Rights Charter (2024/2655(RSP)). Available at https://www.europarl.europa.eu/doceo/document/TA-9-2024-0286_EN.htmlRupčić, 2023, p. 176.

care. If it proves impossible to regulate conscientious objection in a way that respects, protects, and fulfills abortion seekers' rights, conscientious objection in abortion provision may become indefensible.

Concluding Remarks

Despite all these debates, which in Croatia have a rather long and relatively rich history, it remains to be seen what the eventual legal solutions will be on the "problematic issue of abortion." It should also be taken into account that Croatia is a society in which there are opposing views on the prohibition of abortion but also a fair amount of agreement on the value of human life and the need to protect it. Rawls (1996, 243 n. 32) noted that three critical political values should be taken into account in answering this question: the due respect for human life, the ordered reproduction of political society over time, including the family in some form, and, finally, the equality of women as equal citizens. Any reasonable balance of these three values will give a woman a duly qualified right to decide whether to end her pregnancy. This right is recognized today to address the question of how abortion would be performed since surgical abortion due to medical advancement is no longer the only option. The period until which the abortion may be performed, whether it will be allowed only in the first trimester or until the viability of the fetus, although, we must emphasize that the concept of viability is, again, because of medical advancement, a contentious issue. The legislature must reconcile these three values with different understandings of the beginning of life, human dignity, women's human rights, and care for those in the most disadvantaged position during pregnancy. It remains to be seen how much Croatia's membership in the European Union will influence the Croatian legislature, i.e., the recent European Parliament resolutions directly criticizing Croatia as a country where women have difficulty accessing abortion owing to the large number of doctors who use conscientious objection.

References

Alinčić, M. (1976). Law and the status of women in Yugoslavia. *Columbia Human Rights Law Review, 8*(1), 345–372.

Barry, N. P. (1995). *Modern political theory*. Macmillan Press

Bonfiglioli, C., & Žerić. S. (2023). Debating abortion and contraception in socialist Yugoslavia: A microhistorical perspective. *Narodna umjetnost, 60*(1), 147–162. https://doi.org/10.15176/vol60no108

Deneen, P. J. (2018). *Why liberalism failed*. Yale University Press. https://doi.org/10.12987/yale/9780300223446.001.0001

Erdman, J. N. (2023). The WHO abortion care guideline: Law and policy—Past, present, and future. *International Journal of Gynaecology and Obstetrics, 162*(3), 1119–1124. https://doi.org/10.1002/ijgo.15017

Geisler, N. L. (2005). *Kršćanska etika*. Bogoslovni institut.

Hajar, R. (2017). The physician's oath: Historical perspectives. *Heart Views, 18*(4), 154–159.

Hlača, N. (2009). Zakon o "pobačaju" Republike Hrvatske 1978–2008. *Medicina Fluminensis, 45* (2), 142–147.

Holm, S. (2023). The stability of political compromise—Abortion legislation in Denmark and Norway. *Cambridge Quarterly of Healthcare Ethics, 32*(3), 337–343. https://doi.org/10.1017/S0963180122000676

Hrabar, D. (2015). Pravo na pobačaj—pravne i nepravne dvojbe. *Zbornik Pravnog Fakulteta u Zagrebu, 65*(6), 791–831.

Kurjak, A., Maria Carrera, J., McCullough, L. B., & Chervenak, F. A. (2009). The ethical concept of the fetus as a patient and the beginning of human life. *Periodicum Biologorum, 111*(3), 341–348.

Lyerly, A. D., Little, M. O., & Faden, R. R. (2008). A critique of the 'fetus as patient.' *The American Journal of Bioethics: AJOB, 8*(7), 42-W6. https://doi.org/10.1080/15265160802331678

Mapasure, M., & Rakoczy, S. (2016). Afro-katoličke žene u susretu s "Humanae Vitae." *Concilium, 52*(2), 91–107.

Matulić, T., & Balabanić, I. (2019). Selected bioethical questions in the Croatian society according to results of the European values study 2017. *Bogoslovska Smotra, 89*(2), 385–409.

Mladenović, M. 1987). *Osnovi sociologije porodice*. Savremena administracija.

Paul, V. I. (1968). Encyclical Letter Humanae Vitae. *Acta Apostolicae Sedis, 60*, 481–503.

Perinčić, R. (1990). Pobačaj—činjenice i posljedice. *Obnovljeni Život, 45*. (4.), 262–274.

Petrić, N. (1981). *Čovjekove slobode, rađanje, samoupravljanje*. Svjetlost—NIGRO Glas.

Popović, P. (2018). Kritika koncepcije pravednosti usvojene u rješenju Ustavnog suda o tzv. "Zakonu o pobačaju". *Bogoslovska smotra, 88* (1), 131–155.

Rawls, J. (1996). *Political liberalism*. Columbia University Press.

Richards, M. (2004). "Assisted reproduction, genetic technologies, and family life". In J. Scott, J. Treas & M. Richards (Eds.), *The Blackwell companion to the sociology of families*. Blackwell Publishing.

Rittossa, D. (2005). Prijepori o pravu na pobačaj u Republici Hrvatskoj. *Zbornik Pravnog Fakulteta Sveučilišta u Rijeci, 26*(2), 971–997.

Rupčić, D. (2023). Pobačaj pod okriljem seksualnog i reproduktivnog prava i zdravlja žena—pravna analiza Rezolucije Europskog parlamenta 2020/2215 (INI). *Crkva u svijetu, 58*(2), 155–180. https://doi.org/10.34075/cs.58.2.1

Šijanović, S; & Kovačević, T. (2011). Etika u reprodukciji čovjeka. In: Stella Fatović-Ferenčić & Antun Tucak (Eds.), *Medicinska etika*. Medicinska naklada (pp. 167–173).

Śledzińska-Simon, A. (2023). Constitutional framings of the right to abortion: A global view. *International Journal of Constitutional Law, 21*(2), 399–406.

Tucak, I., & Blagojević, A. (2020). Abortion in Europe. *EU and Comparative Law Issues and Challenges Series (ECLIC), 4*, 1135–1174. https://doi.org/10.25234/eclic/11943

Tucak, I., & Blagojević, A. (2023). Rethinking the right to abortion in the United States of America. *Obnovljeni Život, 78*(5), 585–599. https://doi.org/10.31337/oz.78.5.5

Vilić, L., & Tucak, I. (2023). Fetus kao osoba. *Izazovi i perspektive razvoja pravnih sistema u XXI vijeku—Zbornik radova Pravnog fakulteta Univerziteta u Banjoj Luci, 1*(3), 19–38.

Vuletić, N., Ivandić, J., Smajla, N., Klarić, M., & Haller, H. (2020). Medikamentozni Prekid Neželjene Trudnoće. *Liječnički Vjesnik, 142*(7–8), 222–229.

National Regulations, Acts and Court Decisions

Act on Health Measures for Exercising the Right to Free Decision on the Birth of Children, Official Gazette, No. 18/78, 31/86, 47/89 and 88/09.

Constitution of the Republic of Croatia, Official Gazette 56/90, 135/97, 08/98, 113/00, 28/01, 41/01, 55/01, 76/10, 85/10, 05/14.

Convention on the Rights of the Child, General Assembly resolution of United nations, (20. XI. 1989).

Law on Medical Practice, Official Gazette 121/03, 117/087

Ruling of the Constitutional Court of the Republic of Croatia No. U-I-60/1991 et al. of February 21, 2017 and Dissenting Opinion. Retrieved from https://narodne-novine.nn.hr/clanci/sluzbeni/2017_03_25_564.html. Unofficial abridged translation into English available at: https://www.law.utoronto.ca/sites/default/files/documents/reprohealth/croatia_2017_constitutional.pdf

The World Medical Association Declaration of Geneva, Physician's Oath, 1948.

Universal Declaration of Human Rights, United nations General Assemby, Paris, (10. XII. 1948.).

Website References

10th convocation of the Croatian Parliament (22.7.2020)—11th session of the Croatian Parliament—voting results. Retrieved from https://www.sabor.hr/prijedlog-zakona-o-medicinskom-postupku-prekida-trudnoce-%20with-the-final-proposal-of-the-urgent-law-0?t=130671&tid=210323

European Parliamentary Forum on Sexual and Reproductive Rights. 2024. "Contraception Policy Atlas Europe 2024". Retrieved from https://www.epfweb.org/sites/default/files/2024-02/CCInfoEU_A3_EN_2024_FEB12_0.pdf

Hrvatski zavod za javno zdravstvo. 2024. "Pobačaji u zdravstvenim ustanovama u Hrvatskoj 2023. godine". Retrieved from https://www.hzjz.hr/wp-content/uploads/2024/10/HZJZ_-_pobacaji_2023._g..pdf

Platform for reproductive justice—Availability of abortion in the Republic of Croatia. Retrieved from https://www.reproduktivna-pravda.hr/dostupnost-pobacaja-u-republici-hrvatskoj

Pope Francis. 2018. "Catechesi sui Comandamenti, 10/A: Non uccidere". Retrieved from https://www.vatican.va/content/francesco/en/audiences/2018/documents/papa-francesco_20181010_udienza-generale.html

The Draft Law on the Medical Termination of Pregnancy. Retrieved from https://www.sabor.hr/prijedlog-zakona-o-medicinskom-postupku-prekida-trudnoce-s-konacnim-prijedlogom-zakona-hitni-0?t=130671&tid=210323

The Parliamentary Assembly of the Council of Europe and Resolution 2331 (2020) about empowering women ("Promoting Access to Contraception in Europe). Retrieved from https://assembly.coe.int/nw/xml/XRef/Xref-XML2HTML-en.asp?fileid=28675&lang=en

World Health Organization, Abortion Care Guideline, 2022. Retrieved from https://www.who.int/publications/i/item/9789240039483

CHAPTER 9

Gender Refugees and Sex Reassignment Surgery: Medical Tourism from Japan to Thailand

Keisuke Shishido

Abstract The availability of aftercare for patients returning to their native countries is a significant issue associated with medical tourism. Such problems have also arisen regarding sex reassignment surgery (SRS) in Japan. As Thailand is a well-known destination for medical tourism related to SRS, Japanese people travel to Thailand to receive high-quality, inexpensive surgeries. In recent years, there have been reports of people who have undergone SRS overseas and have been denied aftercare in Japanese clinics. This is because the details of the surgeries performed on these individuals are unknown, and aftercare is complex without knowing these details. The inability of these individuals to receive necessary medical care makes them so-called "gender refugees." Despite these risks, one of the reasons that Japanese transgender people visit Thailand is the Japanese legal system. Under Japan's official family registration system, individuals

K. Shishido (✉)
Okayama University Graduate School of Interdisciplinary Science and Engineering in Health Systems, Okayama, Japan
e-mail: kei-shishido@okayama-u.ac.jp

I. Pavić et al. (eds.), *Bioethics and Social Ethics in The Modern World*,
https://doi.org/10.1007/978-3-031-86418-6_9

need an SRS to change their gender status. This study provides background information on such issues and reports on recent notable cases. Finally, comments on the future outlook are provided.

Keywords Medical tourism · Sex reassignment surgery (SRS) · Gender identity disorder (GID)

Background

Currently, in 2024, it is estimated that a significant number of people from Japan[1] will travel to Thailand to receive sex reassignment surgery (SRS). One may point to the unique legal situation in Japan regarding the background of Japanese transgender people seeking SRS.

One of the leading reasons for this is the Japanese family registration system (Koseki). Gender (sex) information is recorded in this official system. Notably, it is our biological sex assigned at birth that is described here and not a person's social gender. In 2003, Japan created a Special Act,[2] the Act on Special Cases in Handling Gender Status for Persons with Gender Identity Disorder (Act No. 111 of 2003), regarding the procedures for changing this official gender record. This Special Act has undergone several amendments, but the term gender identity disorder (GID) in the title has not changed. The term GID is being reviewed internationally from the perspective of depathologization (Nakatsuka 2022). The term "gender dysphoria" is also becoming popular following the DSM-5 published by the American Psychiatric Association in 2013. However, in this study, the term GID is used following the title of the Special Act.

Article 2 of the Special Act defines the GID, and a person requesting a change in sex should meet this definition.

[1] This study was supported by JSPS KAKENHI (grant number: JP22K10480).

[2] An English translation of the Special Act is available from the website of Japanese Ministry of Justice. However, it should be noted that it does not reflect the latest amendments to the Act. See: Ministry of Justice. (n. d). Act on Special Cases in Handling Gender Status for Persons with Gender Identity Disorder. Japanese Law translation. https://www.japaneselawtranslation.go.jp/ja/laws/view/2542.

Article 2: The term "Person with Gender Identity Disorder," as used in this Act, refers to a person whose biological gender is evident but who holds a persistent conviction under which they psychologically identify themselves as being of the opposite gender (hereinafter referred to as the "Opposite Gender"), who has the intention of physically and socially conforming with the Opposite Gender, and who has received concurrent diagnoses on such identification with the Opposite Gender from two or more physicians equipped with the necessary knowledge and experience to give accurate diagnoses on this matter, on the basis of generally accepted medical knowledge.

In addition, they should request that all the requirements of Article 3 (1) of the Special Act be fulfilled. Any person who has undergone a gender reassignment hearing shall be "*unless otherwise provided for by law, concerning the application of the Civil Code (Act No. 89 of 1896) and other laws and regulations, deemed to have been assigned to the Opposite Gender*" (Article 4).

Article 3 (1): A family court may make a ruling of a change in the recognition of the gender status of a person who is a person with gender identity disorder and who falls under all of the following items at the request of such a person:
(i) is not less than 18 years of age;
(ii) is not currently married;
(iii) currently has no child who is a minor;
(iv) has no reproductive glands or whose reproductive glands have permanently lost function; and
(v) has a body that appears to have parts that resemble the genital organs of those of the opposite gender.

Legal and ethical arguments exist for each requirement in Article 3 (1) (Fujito 2020). Those that are relevant to the SRS are items No. (iv) (reproductive inability) and No. (v) (appearance conformity) requirements. Many people undergo SRS to meet these requirements. Consequently, over 10,000 gender status changes have been accepted in Japan over the past 20 years (Gid.jp n.d.).

Few medical institutions in Japan routinely perform SRS. In 2018, the SRS was covered by medical insurance. Initially, this insurance was only available at four medical facilities (Sapporo Medical University, Yamanashi University, Okayama University, and Kousei Hospitals). Many of these medical facilities are far from metropolitan areas (e.g., Tokyo and Osaka) and are inconvenient for parties with GIDs. People with GIDs must worry about high expenses if insurance is unavailable. Therefore, many people travel from Japan to foreign countries where they can receive inexpensive (and sound) SRSs. Thailand is central to SRS medical tourism (Chokrungvaranont et al. 2014).

The situation of SRS medical tourism has been reported in newspapers and the news. Several SRS companies target Japanese people, and their advertisements can be found online. To some extent, such SRS providers have been established as businesses in Japan without official licensing or other legislation. One report covering such companies stated that half of those who operated SRS abroad had undergone SRS in Thailand (Ito 2020). The experts are aware of this situation. For example, Dr. Namba of the Gender Center at Okayama University Hospital stated the following in the abstract of the GID Society's 21st Research Conference (March 23–24, 2019):

> *To date, more than half of domestic GID patients have received SRS in Thailand and other countries, and the reasons for this are the lack of facilities to receive patients in Japan and the high cost of treatment.*

Moreover, it has also become clear that some people who undergo surgery overseas cannot receive follow-up care after their return to Japan (Namba 2014). There have been reports of cases in related societies in which people with GIDs refused medical treatment after returning to Japan. In addition to complications related to SRS, unexpected bleeding or poor conditions can occur occasionally after surgery. Leaving these symptoms untreated is undesirable for their health. "*However, would a physician without experience in SRS accept a patient with such complications?*" (ibid.). The Gender Clinic of Okayama University Hospital accepts patients whom other medical institutions have refused. However, accepting all such patients is becoming increasingly difficult.

It is probably not well known internationally that such "gender refugees" exist, even in Japan, a country with advanced medical care, systems, and technology. We are currently planning to conduct further research on this topic. In addition, a decision that we believe may have had some impact on this issue came from Japan last year. This decision is discussed in the following chapter.

Recent Cases

As mentioned earlier, one of the reasons why Japanese people travel abroad for SRS is legal. The Japanese Special Act, especially Item No. (iv) (reproductive inability) and No. (v) (appearance conformity) requirements, has been criticized. Internationally, in 2014, the WHO issued a joint statement criticizing sterilization as a requirement for legal gender reassignment (WHO, 2014). Additionally, in 2017, the European Court of Human Rights ruled that making surgery or treatment unwanted by the person, as a transgender person, a requirement for legal gender reassignment violates Article 8 of the European Convention on Human Rights (right to respect for private and family life; A.P., Garçon et Nicot c. France du six avril 2017.). The international situation is known to a considerable extent (to experts, if not the general public).

Is surgery required to change the handling of sexual status? The Japanese Supreme Court quickly made two decisions on this issue.

In 2019, the Supreme Court held that requirement No. 4 was constitutional (Supreme Court Judgment January 23, 2019). In this case, a Female to Male (FtM) transgender (X^1) argued as follows and requested a change in the handling of gender status without the operation of Article 3 (1) item (iv) of the Special Act. X^1 argued that Article 3 (1) item (iv) of the Special Act is contrary to the Constitution in the following ways: First, it is contrary to the right not to invade the body guaranteed by Article 13 of the Constitution of Japan; second, it violates Article 14 of the Constitution, which provides equal rights. Article 13 is critical in this case.

> *Article 13: All the people shall be respected as individuals. Their right to life, liberty, and the pursuit of happiness should be the supreme consideration in legislation and other governmental affairs to the extent that they do not interfere with public welfare.*

With respect to the right to pursue happiness, as referred to in Article 13, it is understood that "*individuals have the right to make their own decisions on certain personal matters without interference by public authority*" (Satoh 2020, 212). The Japanese theory of the right to self-determination is derived from this article. Citing these rights, some objected to the enforcement of surgery.

However, the Supreme Court rejected the appeal, stating that "*this provision does not force the above operation for persons with GID in general.*" This decision discouraged LGBTQ parties. Conversely, some believe that the 2019 decision was the Supreme Court's warning to the legislative branch regarding the provision's unconstitutionality (Nakaso 2019). The Supreme Court made a reservation at that time, stating that "*this provision cannot be said to violate Articles 13 and 14(1) of the Constitution at this time.*"

Remarkably, this decision was overturned in less than five short years. In 2023, the Supreme Court ruled that the Special Act's surgical provisions violated the Constitution (Supreme Court Judgment, October 25, 2023). In this case, the plaintiff (X^2) in this 2023 case is a male-to-female (MtF) transgender. X^2 argued that items (iv) and (v) of Article 3 (1) contradict Articles 13 and 14 (1) of the Constitution. As such, the Supreme Court ruled on the requirement of item (iv), reproductive inability. It stated that the requirement for reproductive inability was unconstitutional, in violation of Article 13 of the Japanese Constitution. At that time, the Supreme Court derived "*freedom from invasion of the body*" from Article 13 of the Constitution. The court then held that in light of the "*freedom from bodily invasion,*" the requirement for surgery "*imposes excessive restrictions.*" Moreover, this unconstitutional decision was the unanimous opinion of all 15 Supreme Court judges.

However, in this case, no decision was made regarding the other requirement, namely, appearance conformity. Therefore, the case was sent back to the High Court for reconsideration.

Discussion

At the time of writing (February, 2024), the High Court had not yet ruled on its decision. Nevertheless, the Supreme Court's decision in 2023 has already affected the lower courts.

For example, the plaintiff (X^1) in the 2019 case filed a second petition with the Family Court to change the handling of gender status. In this case, in a previous adjudication, X^1 met the requirement of the Special Act item (v) (appearance conformity) owing to the administration of hormones. Therefore, even without a Special Act amendment or a new decision on the requirement of item (v), the Family Court approved a change in the handling of gender status (Okayama Family Court Tsuyama Branch, Decision, February 7, 2024).

In any case, the situation calls for reconsideration in the Diet regarding the provisions of the special law, but the direction of that reconsideration is unclear. Recent examples of legislation in other countries have already been reported in Japan, in which surgical requirements have been removed in some cases, and other (less invasive) requirements have been put in place in others (Fujito 2020, 94). The dissenting opinions of three judges accompanied the 2023 Supreme Court decision. These dissenting opinions also address the unconstitutionality of the item (v) requirement. However, the 2023 Supreme Court decision itself has not been sufficiently elaborated on in its constitutional judgment (Kawashima 2024).

Furthermore, arguments are being raised against eliminating this surgical requirement in Japan. One of the reasons why appearance conformity was required in the first place was social problems, such as access to segregated public bathhouses (Nohno 2004, 93–94). A group opposes the special Act itself in Japan, advocating the protection of "women's spaces," such as restrooms and public bathhouses. Of course, it is the legislature's role to revise the Act in response to the Supreme Court's decision and to reconcile the various interests involved. Importantly, in Japan, this special act was proposed by members of the Diet (70–90% of Japanese laws are bills submitted by the cabinet). Therefore, it is difficult for the Japanese government to take a proactive stance even if a revision of the Special Act is necessary. Of course, the outcome of the High Court of Appeals, which was remanded, will be important in determining what requirements will change in the future. In addition, other related cases and activities of the parties involved must be closely monitored.

References

Chieko, N. (2004). *Kisetsu seidouitsuseisyougaisya seibetsu toriatsukai tokureihou [Handbook the Act on Special Cases in Handling Gender Status for Persons with Gender Identity Disorder]*. Nihon Kajo Syuppan.

Genki, I. (2020). *Sen ten shi [Sex Reassignment Agent]*. Kashiwa Shobo. (in Japanese)

Gid.jp (n.d.) *Seidouitsuseisyougai tokureihou ni yoru seibetsu no toriatsukaino hennkousuu chousa [Survey on the Number of Changes in handling gender status under the Special Act on Gender Identity Disorder (2022)]*. Japan Association of People Living with Gender Identity Disorder and Gender Dysphoria. Retrieved from https://gid.jp/research/research0001/research2023101001/ (in Japanese)

Haruna, K. (2024). Saikousai ni yoru seidouitsuseisyougai tokureihou no iken kettei [Unconstitutional Decision by the Supreme Court on the Gender Identity Disorder Special Act]. *Hougaku Seminar, 829*, 40–46. (in Japanese).

Hisao, N. (2019). Hanrei kenkyu: Seidouitsuseisyougai Tokureihou 3 jou 1kou 4 gou no goukensei (saikousai kettei heisei 31 nen 1 gatsu 23 nichi) [Case Study: Constitutionality of Article 3 (1) item (iv) of the Special Act for Gender Identity Disorder (Supreme Court Decision, January 23, 2019)]. *Research for Regional Collaboration, 1*, 37–42. (in Japanese).

Kouji S. (2020). *Nihonkoku kenpou ron [Constitutional Law of Japan (2nd ed.)].* Seibundoh. (in Japanese)

Mikiya, N. (2022). Sei no tayousei—seidouitsuseisyougai/seibetsufuwa LGBTQ/SOGI toha? [Gender Diversity—What is Gender Identity Disorder/Gender Dysphoria, LGBTQ/SOGI?]. *Japanese Journal of Pediatric Medicine, 54*(10), 1672–1676. (in Japanese).

Prayuth, C., et al. (2014). The development of sex reassignment surgery in Thailand: A social perspective. *The Scientific World Journal*, 1–5. https://doi.org/10.1155/2014/182981

WHO. (2014). *Eliminating forced, coercive and otherwise involuntary sterilization: An interagency statement.* http://apps.who.int/iris/bitstream/10665/112848/1/9789241507325_eng.pdf

Yoshitaka, F. (2020). Houteki seibetsu henkou ni kansuru Nihon oyobi syogaikoku no hou seido [Laws on Gender Recognition in Japan and other Countries]. *Research and Legislative Reference Bureau National Diet Library, 830*, 79–100. (in Japanese).

Yuzaburo, N. (2014). Comprehensive treatment for gender identity disorder. *Journal of Okayama Medical Association* 126(2), 109–125. (in Japanese). https://doi.org/10.4044/joma.126.109

CHAPTER 10

Using Traditional, Complementary, and Alternative Medicine Amidst the COVID-19 Pandemic: Results of a Database Search Focusing on Diverse Approaches and Usage

Yutaka Kato

Abstract The emergence of the COVID-19 pandemic has motivated many people to look for complementary and alternative medicine (CAM) approaches, in order to prevent the infection or treat the disease. In this chapter, CAM practices during the pandemic are analyzed, with the focus on the categorization of CAM approaches and their ethical consequences. By analyzing the PubMed literature, the following CAM approaches were identified: beneficial substances (tested for their effectiveness), bodily manipulation, mind–body medicine, alternative medical systems, energy therapy, etc. Discussion of the literature identifies relevant ethical issues (differentiating between complementary and alternative uses, establishing

Y. Kato (✉)
Shiga University of Medical Science, Otsu, Shiga, Japan
e-mail: yk227@belle.shiga-med.ac.jp

I. Pavić et al. (eds.), *Bioethics and Social Ethics in The Modern World*,
https://doi.org/10.1007/978-3-031-86418-6_10

evidence for CAM effectiveness, potential biases in CAM research, and conflicts of interest among CAM practitioners). An additional emphasis is placed on understanding the differences between the complementary use of CAM and its integration into health policy.

Keywords COVID-19 · Complementary and alternative medicine (CAM) · Efficacy and safety · Ethical implications

Introduction

The novel coronavirus pandemic has posed a significant threat since its onset in 2019. The gravity and urgency of the situation encouraged some people to explore or utilize traditional, complementary, and/or alternative medicine approaches/methods. (The terms traditional and complementary medicine (T&CM) are also utilized. Nevertheless, since traditional approaches are often used in either a complementary or alternative manner, the term complementary and alternative medicine (CAM) is henceforth employed.) Over-the-counter Kampo drugs have also been utilized by the general public in Japan. Kampo represents a unique adaptation of traditional Chinese medicine introduced to Japan. This phenomenon is understandable because it is common to use *kakkonto* to treat the common cold, some of which may be caused by coronavirus (not the novel coronavirus). In 2022, Kracie Pharmaceutical, Ltd., a pharmaceutical company headquartered in Tokyo, designated *bakumondoto* the Kampo of the Year. *Bakumondoto* is considered effective against respiratory tract infections and symptoms such as coughing.[1] This phenomenon of resorting to traditional and alternative medicine approaches is presumably a worldwide occurrence. In principle, agencies of the United Nations have promoted traditional medicine.[2] A database search was conducted

[1] Kracie, Ltd. "KAMPO OF THE YEAR 2022. *Bakumondoto* Emerges as the Most Notable Kampo Drug. Demand for Kampo Medicines Among Young People in Their 20s Has Been Increasing Over the Last Five Years." December 1, 2022. https://www.kracie.co.jp/release/10180959_3833.html.

[2] See, e.g.: "WHO traditional medicine strategy: 2014–2023." May 15, 2013. https://www.who.int/publications/i/item/9789241506096.

to gain an overview of the approaches researched worldwide during the pandemic and explore relevant ethical implications.

Methods

The database search was conducted using the PubMed database with the following terms: ((Traditional Medicine COVID-19[Title/Abstract]) OR (Complementary Medicine COVID-19[Title/Abstract])) OR (Alternative Medicine COVID-19 [Title/Abstract]) OR (CAM COVID-19 [Title/Abstract]). The data explored was restricted to publications available as of November 2, 2022, and published since February 1, 2020. The search yielded 542 results. Among the initial results, 447 articles were excluded due to their irrelevance based on the article titles, and 53 were excluded on the basis of the content of the abstracts. Some articles reported solely on research conducted during the COVID-19 pandemic and were therefore excluded. Only articles addressing conditions caused by the COVID-19 were included ($N = 42$).[3]

Results of the Database Search

The relevant characteristics of the identified bibliometric records are geographical distribution, journals, conflicts of interest or competing interests, diverse approaches, effectiveness, adverse effects, evidence, purposes, and complementary or alternative usage. This chapter discusses diverse approaches and complementary or alternative usage.

Diverse Approaches

The approaches identified in the search results were diverse. To grasp this diversity, the approaches are presented below according to the categorization proposed by the National Center for Complementary and Integrative Health (NCCIH) Classification System[4]:

[3] The PMIDs of the articles are provided in the Appendix.

[4] Available at https://www.sciencedirect.com/topics/medicine-and-dentistry/national-center-for-complementary-and-integrative-health. There are several other taxonomies available to classify the CAM approaches.

1. "*biologically based systems which consist of herbal supplements, therapeutic foods, and overall diet*": garlic, ginger, lemon, *Tinospora cordifolia* ("a widely grown shrub" from India) (used in Ayurvedic medicine), curcumin, *Shunthi* (*Zingiber officinale* (ginger)), Vidanga (*Embelia ribes*), *Yashtimadhu* (*Glycyrrhiza glabra*), *Haritaki* (*Terminalia chebula*), Shatavari (*Asparagus racemosus*), *Aamalaki* (*Emblica officinalis*), *Pippali* (*Piper longum*), minerals, vitamins, calcined zinc, Siddha medicines, Kaba Sura Kudineer (KSK) & Nilavembu Kudineer (NVK), phytotherapy, hydrogen peroxide;
2. "*manipulative and body-based systems which include massage, chiropractic, and osteopathic medicine*": acupuncture, warm cupping;
3. "*mind–body medicine spanning the practices of yoga, tai chi, and medication*": yoga, mindfulness meditation, relaxation
4. "*alternative medical systems which include homeopathy, naturopathy, traditional Chinese medicine, and Ayurveda*": Ayurveda, *Shankha bhasma* (an Ayurvedic medicine), Chinese herbal medicine, traditional Chinese medicine, (Siddha medicine), naturopathy, homeopathic remedies, and *Arsenicum Album* (a homeopathic solution prepared by diluting aqueous arsenic trioxide);
5. "*energy therapy consisting of Reiki, magnet therapy, therapeutic touch, and qi gong*": self-Reiki, qigong.

The relationships between each category are not mutually exclusive. For example, Category (1) includes those utilized in category (4) Ayurveda and Siddha medicine. (Especially in category (4), a single item may encompass complex, systematic, and varied procedures.) Thus, an approach may fall into multiple categories. That said, it is evident that the approaches span all categories.

The inclusion of numerous items in category (1) suggests that researchers have sought to identify beneficial substances. This is not a new phenomenon. For example, acetylsalicylic acid and artemisinin were used as traditional remedies before being industrially manufactured and marketed as modern medicines.

Also, it is important that the terms were used in a somewhat confused manner. Specifically, only "complementary" was utilized in articles numbered 165, 215, 326, 497, and 542 (and inconsistently in 51). In other articles, "complementary" consistently appeared together with "alternative."

Discussion of Ethical Implications

Many of the approaches that were already known were found in the search results. In particular, the symptoms and sequelae of COVID-19 vary, and prevention methods have been explored. Thus, studies of CAM approaches have targeted these diverse areas. Practitioners would have attempted to address COVID-19 via the approaches they routinely employ. Every practice within the taxonomy has likely been explored over time.

Generally, CAM approaches are not entirely separate from modern conventional medicine. For example, willow is traditionally used for pain relief, from which salicylic acid is extracted and acetylated, yielding acetylsalicylic acid, which is still widely used today. Additionally, Mao, a herb containing ephedrine—a substance used in modern medicine, is utilized in popular Kampo formulas such as *kakkonto*, *shoseiryuto*, and *maoto*. Although not included in the results of the search, over-the-counter Kampo drugs are commonly used by the general public in Japan, as mentioned in the introduction section. (Traditional Chinese medicine was studied in articles 406 and 498). Medical doctors in Japan can prescribe both Western conventional medicines and certain Kampo medicines together, and both are covered by health insurance, allowing for a wider range of treatments. Thus, certain Kampo medicines, though traditional, are part of Japan's health policy.

In Japan, it is sometimes casually stated that traditional medicine results from thousands of years of human experimentation (i.e., trials and errors). However, despite its long history of observation, determining the precise health benefits without rigorous measurements remains challenging. Evaluating the efficacy of a drug becomes particularly difficult if it primarily enhances general health and immunity rather than producing the acute effects typical of modern industrially produced drugs. Many papers have reported insufficient evidence to support the efficacy of CAM approaches for treating COVID-19. Establishing evidence, not only for emerging infectious diseases, is time-consuming. Given the rate of evidence accumulation reported in the literature, immediate evidence-based responses to similar pandemics via CAM approaches may not be feasible.

Publication bias, the tendency to favor the publication of studies with statistically significant results over those with nonsignificant results, often leads to a greater proportion of statistically significant studies in academic journals. Nevertheless, demonstrating the inefficacy of commonly practiced methods is also crucial. Given the widespread interest in the effectiveness of specific procedures in treating COVID-19, studies with statistically insignificant or adverse outcomes should also be published.

Nevertheless, CAM approaches present a unique challenge in achieving a balanced representation of studies. Practitioners of these approaches often lack incentives to publicize negative results and may be motivated to conceal adverse outcomes. This conflict of interest may also be observed among pharmaceutical companies and medical professionals, but they can explore other interventions if one fails. In contrast, CAM practitioners, such as those practicing tai chi, qigong, siddha medicine, reiki, etc., may find it more difficult to abandon their practice and adopt an entirely new approach.

While some articles may appear biased and lack scientific rigor, proving their nature definitively can be challenging.[5] Even for approaches that seem unscientific, a search for online resources reveals a multitude of arguments in their defense, making it unlikely to reach a universally convincing decision. When many are invested in a particular viewpoint, there is an incentive to generate discourse that supports their position.

Moreover, the concept of EBM opens the possibility that procedures whose mechanisms are unknown will be utilized, prioritizing benefits and effectiveness in public health.[6] For one thing, if a placebo offers relief, resorting to a placebo may be preferable to enduring symptoms untreated. However, concerns arise regarding the allocation of resources toward a mere placebo. In the UK, it has been determined that public funds should not be allocated to homeopathy, as they are deemed no more effective than a placebo.[7] While reliance on placebos is undesirable when effective

[5] For example, the Japanese Ministry of Health, Labor and Welfare hosts a website offering information on integrative medicine. On this platform, the Ministry translates information provided by NCCIH, which denies the effectiveness of Reiki. It is doubtful that such information will dissuade its followers. (The website does not mention Covid-19).

[6] Achilleas Thoma and Felmont F. Eaves. "A Brief History of Evidence-Based Medicine (EBM) and the Contributions of Dr David Sackett." Aesthetic Surgery Journal, Volume 35, Issue 8, November/December 2015, pp. NP261–NP263. July 10, 2015. https://academic.oup.com/asj/article/35/8/NP261/251339.

[7] Laura Donnelly. "NHS to ban homeopathy and herbal medicine, as 'misuse of resources.'" July 21, 2017. The Telegraph. https://www.telegraph.co.uk/news/2017/07/21/nhs-ban-homeopathy-herbal-medicine-misuse-resources/

interventions are available, there may be instances where it is justifiable to allow people to utilize placebo treatments or their equivalents in the absence of alternative measures. The amount that can be justifiably spent on such measures depends on the severity of the situation and other relevant factors.

None of the articles discussed the problem inherent in choosing CAM over standard measures. However, according to one article (132), individuals who used CAM tended to have a lower inclination to get vaccinated, indicating potential issues linked to the alternative use of CAM approaches.

The distinction between alternative and complementary use has significant implications. (Some confusion was observed in the literature.) The US Congress renamed the National Center for Complementary and Alternative Medicine to the National Center for Complementary and Integrative Health (NCCIH) in December 2014.[8] Instead of using "alternative," which is no longer favored, the NCCIH now employs "integrative," and its mission is to investigate "complementary and integrative health approaches" (https://www.nccih.nih.gov/about). Using "alternative" approaches is more problematic, as it typically implies the absence or denial of standard measures. Conversely, when complementary, these approaches are less problematic because standard procedures are not excluded. Nonetheless, in the case of complementary use, one must be cautious about interactions and monitor whether the combination causes adverse effects or inhibits positive effects.

Additionally, in the case of the COVID-19 pandemic, whether specific procedures are used as alternative or complementary measures is important. However, as shown above, this distinction has not been made in many studies. Since a particular procedure can be utilized either way, it is understandable that research is being conducted with reservations about whether the procedure in question should be used as an alternative or complementary measure. At the very least, a clear distinction should be made when these procedures are implemented. One focus of research should be determining how effectively specific approaches can be used complementarily.

There are specific procedures that, by nature, cannot be used simultaneously (e.g., continued dialysis would be unnecessary after a kidney

[8] "NCCIH Timeline." https://www.nccih.nih.gov/about/nccih-timeline.

transplant). However, traditional medicine approaches may be used concurrently with modern medical interventions.[9]

However, when financial costs are considered, alternative use may seem more justifiable. To illustrate this point, acupuncture/moxibustion can be alternatives in Japan. These procedures are covered by insurance only when conventional medical interventions have proven ineffective, but patients are not required to seek CAM covered by insurance.[10] For example, the rehabilitation center at the University of Tokyo Hospital, one of the most prestigious hospitals in the country, offers acupuncture and moxibustion only outside of health insurance, and patients are not eligible for these treatments on the same day as insurance-covered procedures—a system designed to prevent excessive treatments.[11]

In the case of emerging infectious diseases, it is challenging to ascertain whether conventional medical interventions are ineffective and meet the aforementioned conditions where CAM procedures are covered by health insurance. That said, while the complementary use of CAM approaches seemed more justifiable, achieving a combination of or balance between

[9] Mokichi Okada (1882-1955) and his followers employed a Reiki-like approach called tekazashi (holding the hand of a practitioner over a patient). Okada held his own unique view of medicine. In general, every drug has side effects. However, Okada's theory of drug toxicity extended beyond this assertion. He believed that disease is the suffering associated with the process of eliminating toxins from the body that are not originally present. According to him, drugs interfere with this process, and the drugs themselves are toxins that do not naturally exist in the body (Yoshihiko Sugioka. Human Beings Suffering Together - Revisiting Religion and Medicine from the Philosophy of Medicine. Shunjusha. February 2023, p. 283). Okada expressed confidence in the Johrei treatment, stating that it was "dozens of times" more effective than modern medicine. Although he was willing to accept verification by modern science, he actually refused to test his technique in an interview with an announcer because he was busy educating his students (Sugioka 294). According to his followers, he used Johrei in conjunction with medical treatment, including taking medicine and undergoing surgery (Sugioka 295).

[10] Acupuncture, moxibustion, and anma massage may be covered by health insurance only when treatment at a medical institution for a specified period of time has been ineffective, and the physician approves the patient, with written consent, to receive acupuncture, moxibustion, and anma massage treatment. Injuries and illnesses that can be covered by acupuncture and moxibustion are limited to neuralgia, rheumatism, cervico-omo-brachial syndrome, frozen shoulder, lumbago, sequelae of cervical sprain, etc. Written consent is required every six months. Tokyo Acupuncture & Moxibustion Association. "Treatment with insurance." https://harikyu-tokyo.or.jp/general/insurance.

[11] The University of Tokyo Hospital. "Rehabilitation Medicine." https://www.h.u-tokyo.ac.jp/english/centers-services/clinical-divisions/rehabilitation-medicine/index.html.

the complementary and alternative use of CAM would be the optimal approach.

Conclusions and Limitations of the Study

The database search identified various CAM approaches targeting the symptoms and prevention of COVID-19. Research on these approaches has varied from plausible to implausible. The distinction between alternative and complementary use of CAM approaches has significant implications, as seen in the shift in terminology from "alternative" to "integrative" by the NCCIH; however, confusion persists in the literature regarding this differentiation.Whereas the complementary use of CAM approaches seems generally more justifiable, achieving a combination or balance between the complementary and alternative use of CAM would be the optimal approach.

Articles that did not use any of the keywords, although they analyzed the subjects considered CAM, were not included in the study. Additionally, this study did not utilize the term "integrative" as a search term. Some time has passed since the database search was conducted for this study. New situations may have arisen since then. The fact that mainstream strains have been replaced one after another has also made it difficult to accumulate evidence, but some evidence may have been established since then.

Acknowledgements This study was funded by JSPS KAKENHI Grant 21K10325.

Appendix: PubMed Search Results

1: PMID: 36277303, 2: PMID:36245699, 30: PMID:34993740, 36: PMID:36213650, 50: PMID: 36057614, 51: PMID: 36141290, 60: PMID: 35308033, 84: PMID: 35966751, 112: PMID: 35677618, 122: PMID: 35770550, 127: PMID: 35747116, 132: PMID: 35736974, 165: PMID: 35427237, 201: PMID: 34826590, 206: PMID: 33992585, 212: PMID: 35193642, 215: PMID: 35168630, 245: PMID: 35398830, 249: PMID: 35002106, 256: PMID: 34840104, 263: PMID: 34462625, 265: PMID: 34230788, 282: PMID: 34777985, 289: PMID: 34395188, 298: PMID: 34820152, 301: PMID: 34805649, 307: PMID: 34230786, 309:

PMID: 33962011, 326: PMID: 34325280, 334: PMID: 33838089, 354: PMID: 34577790, 373: PMID: 34335827, 390: PMID: 35095365, 394: PMID: 34050125, 406: PMID: 33761591, 416: PMID: 33832446, 439: PMID: 33573696, 461: PMID: 33162692; 480: PMID: 33225970, 497: PMID: 33066867, 498: PMID: 32792254, 542: PMID: 32250655.

References

Bendezu-Quispe, G., Benites-Meza, J. K., Urrunaga-Pastor, D., Herrera-Añazco, P., Uyen-Cateriano, A., et al. (2022). Consumption of herbal supplements or homeopathic remedies to prevent COVID-19 and intention of vaccination for COVID-19 in Latin America and the Caribbean. *Tropical Medicine and Infectious Disease, 7*(6), 95. https://doi.org/10.3390/tropicalmed7060095

Ergin, E., & Ozbolat, G. (2022). Nurses' use of holistic, complementary, and alternative medicine methods against COVID-19 anxiety. *Alternative Therapies in Health and Medicine,* 28(3), AT7206.

Kladar, N., et al. (2022). Phytotherapy and dietotherapy of COVID-19—An online survey results from the central part of the Balkan Peninsula. *Healthcare, 10*(9), 1678. https://doi.org/10.3390/healthcare10091678

Kathiresan, N., et al. (2021). "It is the best part of our hospital life": A qualitative analysis on the impact of yoga and naturopathy as a complementary therapy in the management of COVID-19. *Asian Journal of Psychiatry, 64*, 102789. https://doi.org/10.1016/j.ajp.2021.102789

Konakci, G., et al. (2020). In the Turkish news: Coronavirus and alternative & complementary medicine methods. *Complementary Therapies in Medicine, 53*, 102545. https://doi.org/10.1016/j.ctim.2020.102545

Mulder, L. T. C., et al. (2022). Prevalence and predictive factors of complementary medicine use during the first wave of the COVID-19 pandemic of 2020 in the Netherlands. *BMC Complementary Medicine and Therapies, 22*(1), 43. https://doi.org/10.1186/s12906-022-03528-x

Weeks, J. (2020). Call to action: announcing the traditional, complementary and integrative health and medicine COVID-19 support registry. *Journal of Alternative and Complementary Medicine, 26*(4), 256–258. https://doi.org/10.1089/acm.2020.29083.jjw

CHAPTER 11

Green Consumer Ethics and Values—Which Role Does the Media Play in Presenting and Shaping Individual Opinions of Ethics and Sustainability?

Isabell Koinig

Abstract In recent years, corporate social responsibility (CSR) has begun to resonate with consumers worldwide, increasing the value of green consumption (Dikici et al., 2022). According to a recent IBM report, *"more consumers [are] driven by purpose than by value"* (IBM, 2022). This suggests that in addition to economic reasoning, which prioritizes profit, companies must also consider economic and social matters (see the triple bottom line by Elkington, 1994), as investments in ethical issues seem to pay off in the long run (Hassan et al., 2022). The communication scholar Marshall McLuhan postulated early on that people base their knowledge on the information they encounter in media articles and broadcasts (Luhmann, 1995). Nevertheless, the media also plays an essential role

I. Koinig (✉)
Department of Media and Communications, University of Klagenfurt, Klagenfurt Am Wörthersee, Klagenfurt, Austria
e-mail: isabelle.koinig@aau.at

I. Pavić et al. (eds.), *Bioethics and Social Ethics in The Modern World*,
https://doi.org/10.1007/978-3-031-86418-6_11

in how individuals perceive and respond to different topics. The media identify and publicize issues that receive public attention (agenda-setting theory; McCombs, 2005). The media steer individual thoughts in a particular direction by emphasizing selected arguments in their reporting (priming; Hoewe, 2020). Finally, the media structure how individuals think about topics (framing; Carter, 2013). The selected theories perceive the consumer as the sole recipient of mass media messages. However, the chapter concludes with an outlook that takes up the concept of the *prosumer*—a view in which individuals are no longer at the mercy of media companies but are enabled to produce their content actively. While social media renders consumers' voices more powerful, stemming from its potential to connect with like-minded individuals throughout the world (Gummerus et al., 2017), green and sustainable consumers might even consider engaging in activism (e.g., including boycotts or bans) to force companies to rethink their business practices (Wirt et al., 2017).

Keywords Green values · Sustainability · Ethics · Media effects · Agenda setting · Priming · Framing · Social media · Prosumers

Introduction

Against the backdrop of more significant social developments, including environmental movements, such as *Fridays for Future* or *Scientists for Future,* which pressure individuals and governments alike to address larger social issues (Dikici et al., 2022), consumer ethics have undergone some dramatic changes (Wang & Kuah, 2018). Defined as *"the moral principles and standards that guide the behavior of individuals as they obtain, use, and dispose of goods and services"* (Muncy & Vittell, 1992: 298), consumer action is seen to rise in relevance globally, given its implications for business operations, the environment, and society at large (Carrington et al., 2020; Vittell, 2015). However, it needs to be perceived as a double-edged sword, as, on the one hand, consumers' unethical behavior can directly affect a firm's bottom line, whereas, on the other hand, ethical action can be seen to encourage companies to reconsider their mode of operation (Dikici et al., 2022; Geiger-Oneto & Arnould, 2011).

In recent years, corporate social responsibility (CSR) has begun to resonate with consumers worldwide, increasing the value of *green consumption* (Dikici et al., 2022). These values suggest that consumers express concern for environmental issues through their purchasing and consumption behaviors (Dikici et al., 2022; Haws et al., 2014). In part, these values are expressive of consumers' characteristics (Cleveland et al., 2005; Lu et al., 2015; Mutunm et al., 2021), whereas, on the other hand, they might also be expressive of individuals' ethics (Halder et al., 2020) or changing consumption preferences (Yan et al., 2021). Green consumption behavior is conceptualized as individuals reducing the strain on the environment and society at large through their consumption choices (Chan, 2001; Nguyen et al., 2018; Uddin & Khan, 2018).

Companies have started to invest in CSR to help consumers change their behaviors. CSR generally refers to the *"responsibility to meet the economic, legal, ethical, and discretionary expectations that society has of corporations at a given time"* (Carroll, 1999). Companies worldwide have adopted CSR, which is seen as a means of establishing legitimacy and is perceived as "a positive normative evaluation of the organization and its activities" (Colleoni, 2013: 228).

CSR usually requires an organization to communicate with purpose. In fact, *"the power of purpose is driving consumer preference and loyalty in a world where trust in corporations is low, and differentiation between brands is negligible"* (Goodpurpose, 2012). On an elementary level, CSR communication refers to all communicative attempts regarding CSR, whereby corporate and promotional communication activities are employed to inform the public about CSR initiatives and projects, and also intend to strengthen the company's corporate social image (Golob et al., 2013). As such, CSR has entered the public discourse, where it, to date, is a relatively marginal issue (Weder, 2012) despite its potential to rise to a *"common sense issue"* (Weder, 2017).

Changing Consumer Values and Ethics: Moving Toward "Greener" Values

The increasing interest of consumers in CSR engagements is confirmed by a recent IBM report, which revealed that *"more consumers [are] driven by purpose than by value"* (IBM, 2022). This suggests that in addition to economic reasoning, which prioritizes profit, companies must also consider economic and social matters, as investments in ethical issues seem to pay off in the long run (Hassan et al., 2022).

Whether this claim indeed holds true is subject to debate, as an article published in the Harvard Business Review found the "green consumer" to be an "elusive" concept (White et al., 2019). The authors claim that while CSR investments have been found to resonate with consumers, their willingness to pay a premium remains low (White et al., 2019). These findings align with previous results (Biswas & Roy, 2015; Chen, 2020; Dorrn & Verhoef, 2015).

These findings are thus also reflective of an attitude–behavior gap identified by Carrington et al. (2014, 2016) and Joshi and Rahman (2015), who reported that despite consumers' general willingness to purchase green products or adopt an economically friendly lifestyle, they fail to do so (Chekima et al., 2017; Echegaray & Hansstein, 2017; Joshi & Rahman, 2015; Mamun et al., 2018). Hence, the following question remains: how can consumers be convinced to follow their initial inclinations? One possible way of creating awareness of the necessity of sustainable and ethical consumer behavior might be by changing the media discourse accordingly.

Traditional View: Media's Potential in Shaping Individual Responses

The communication scholar Marshall McLuhan already postulated that people base their knowledge in large parts on the information they encounter in the media, which may be in the form of news articles, radio or television broadcasts, or, in the digital age, online news outlets, blogs, or podcasts (Luhmann, 1995). This is a rather traditional view of mass media, according to which the media is used to induce individuals to think about selected topics, also guiding how they should perceive these topics (Cohen, 1963; Scheufele, 2000).

In this context, three communication theories might offer valuable insights.

Agenda Setting

The media identify and publicize issues that receive increased public attention (*agenda-setting theory*; McCombs, 2005). When individuals cannot make sense of the world by themselves, they turn to the media, which presents them with content on which to base their social evaluation (as part of *cultivation theory*; Bilandzic & Rössler, 2004). News reporting, however, is selective, and individuals have come to rely on the media to pre-select relevant topics (Hansen, 2016). Agenda setting describes *"the importance assigned to issues and personalities in the media"* (Winter & Eyal, 1981: 376). Building on the assumption that media (executives) emphasize selected topics or stories over others, e.g., by covering them in more detail, neglecting or entirely omitting others (Price & Tewksbury, 2000), or placing stories more prominently, attention is given to selected topics, which then become excessively debated (McCombs & Shaw, 1972). The discourse of sustainability and ethical consumption thus benefits from news media reporting on relevant topics, such as sustainable production or recycling, which can positively impact individuals' intentions to model such behavior.

Three levels of agenda setting need to be distinguished. The *first level of agenda setting (basic agenda setting)* refers to the prominence or attention topics or issues receive as part of the public discourse (McCombs et al., 2014). While the media might not induce individuals to think in a particular way, it might be able to select the topics with which individuals concern themselves (Cohen, 1963), as demonstrated by a BBC article reporting on ethical business and consumption trends (BBC, 2019). This phenomenon has become known as (issue) salience (McCombs et al., 1997) or message prominence (Arguente et al., 2007). The *second level of agenda setting* or *attention agenda setting* describes which attributes individuals ascribe to topics or objects, thus steering public opinion toward certain viewpoints. Cristianne Close's (2021) article for the World Economic Forum discusses eco-awakening as a viable trend among consumers and can be read in support of sustainable consumption. Other articles, however, might take a more antagonistic stance (e.g., Mongabay, 2023). The *third level of agenda setting*, also known as *network agenda setting*, concerns how issues (first level) and attributes (second level) are connected to construct perceptions of social reality (McCombs et al., 2014). For example, Mazza et al. (2022) identified 12 interrelated topics that dominate the media discourse of energy corporations on Twitter. These perceptions are sometimes subject to individuals' sociodemographics, as Lillo-Viedma et al. (2022) established through their network analysis.

Priming

The media steer individual thoughts in a particular direction by emphasizing selected arguments in their reporting (*priming*; Hoewe, 2020). Specifically, priming builds on or even extends agenda setting by suggesting that individuals' evaluations are informed by thematic issues addressed as part of media coverage (Iyengar & Kinder, 1987; Scheufele, 2000). While initially associated with political communication and political priming (Scheufele, 2000), priming can also be applied to sustainability, as the media offer individuals viable information on pro-social and environmentally friendly behavior (Schulz, 2003). The media provides the general public with information on the severity of climate change, environmental degradation, and pollution, thus allowing individuals to evaluate their role in rectifying the environmental damage done to the planet.

During priming, previous reactions to stimuli or events are taken as reference points to predict individual behaviors when individuals are exposed to a subsequent stimulus or event (Roskos-Ewoldsen et al., 2007, 2009). As one of the core socializing agents, the media uses desirable behaviors that individuals should model, and which are reemphasized in media coverage (Paus-Hasebrink et al., 2019). Through priming, i.e., emphasizing selected norms and favorable outcomes, the media can drive *"changes in the standards that people use to make [...] evaluations"* (Iyengar & Kinder, 1987: 63). Therefore, recent primes are more readily activated and recalled—an effect that has become known as recency (Higgins et al., 1985). Similarly, more potent primes persist over time (Higgins et al., 1985). Priming has been found to predict short-term effects in individuals who have been exposed to violent or political media content (Roskos-Ewoldsen et al., 2007; Ewoldsen & Rhodes, 2020). For sustainability and ethical consumption, this suggests that if the media favorably portrays these trends, it can, for example, influence how individuals perceive these topics, ultimately shaping individuals' behaviors and mindsets (Wang & Anderson, 2008; Fatemi & Dube, 2021).

When messages are primed, message salience increases (Scheufele, 2000). Salience cues assist individuals in establishing the importance attributed to an issue or topic (Wanta & Tarasevich, 2019). These could, for example, be linguistic cues or visual cues, the prior alluding to the status assigned to the topic, whereas the latter finds expression in visual form, e.g., through pointers or variations in color or form (Burmerster et al., 2018). In their study, Tanford et al. (2020) reported that affective priming influences individuals' evaluations and behavioral intentions, resulting in them taking pro-environmental action.

Framing

Finally, media structure how individuals think about topics (*framing*; Carter, 2013; for an overview, see also Moy et al., 2016). Extending agenda setting (Scheufele, 2000), framing refers to "the selection of a restricted number of thematically related attributes for inclusion on the media agenda when a particular object is discussed" (McCombs, 1997: 6). According to Pan and Kosicki (1993), frames need to be perceived as both journalistic norms and organizational statutes upon which content presentation is based (Gamson & Modigliani, 1987).

Frames are specific "modes of presentation" (Scheufele & Tewksbury, 2007: 12) and are helpful in that they assist individuals in structuring and processing information (Fischer & Johnson, 1986). As such, they present interpretative "primary frameworks" (Goffman, 1974: 24; see also Scheufele, 1999 or Price & Tewksbury, 2000) that are, on a macro level, commonly used to reduce the complexity of issues by activating existing schemas (Gans, 1979; Shoemaker & Reese, 1995; Gitlin, 1980). In a systematic review, loss framing (i.e., messages emphasizing what individuals would lose when not performing a recommended behavior) was confirmed to be more effective than gain framing (i.e., messages emphasizing what individuals would gain when performing a behavior) in convincing individuals to adopt a green behavior (Homar & Cvelbar, 2021).

In general, natural and societal frames need to be distinguished, whereby the former occur naturally and unintentionally, helping individuals make sense of the information they are presented with. For example, if individuals are knowledgeable about sustainable consumption or energy preservation, media messages that align with their values and predispositions are more readily understood. Societal frames, on the other hand, are created intentionally, assisting individuals in forming specific opinions and ideas that conform with social norms (Goffman, 1974). Here, media messages emphasizing the need for recycling or implementing a law to reduce greenhouse gas emissions might promote desirable social behaviors or norms (Euronews, 2023). Moreover, positively framed (*gain-framed*) media messages are more likely to induce individuals to change their behavior (Tanford et al., 2020).

More Contemporary View: Individuals as Active Recipients and Prosumers of Media Content

The theories presented above perceive the individual to be passive, i.e., the sole recipient of mass media messages. However, in the present-day media environment, individuals are no longer at the mercy of media companies. Instead, they can produce content themselves and proactively participate in the media discourse (McCombs & Valenzuela, 2020). In the new hybrid media environment, social media users are producers and consumers, rendering their voices more powerful (Bruns, 2007), stemming from their potential to connect with like-minded individuals worldwide (Gummerus et al., 2017). Conditioned by their changed role, users can even engage in grassroots activism, raising topics to the media agenda that might otherwise not have received consideration (Skaidré, 2019).

Consequently, it is often claimed that the media's power in shaping individuals' worldviews is currently less pronounced (Nowak, 2016), given the plethora of channels available for individuals to consult. Some researchers even plead for a differentiation between media actors, grouping them into traditional offline media, traditional online media, and online social networks (Weimann & Brosius, 2015). In the process, new actors have entered the field, with bloggers or influencers acting as mediators (Weimann et al., 2014), who might negate the media's gatekeeping role by reframing topics encountered in traditional media (Weimann & Brosius, 2015). In doing so, they assume a *gatewatching* function (Nowak, 2016).

Exploiting the potential of both their new roles and those of new media infrastructures, consumers are awarded a platform to voice their opinions or induce companies to change their policies and modes of operation (Wirt et al., 2017; Zeng et al., 2020). The two social and environmental movements mentioned in the introduction, *Fridays for Future* and *Scientists for Future,* originated from individuals connecting over shared interests. Despite being based locally, both groups managed to attract members worldwide, relying on social networking sites to grow their member base. For example, *Fridays for Future* was formed in the summer of 2018, and more than 6 million people participated in its greatest strike on September 27, 2019 (Taylor et al., 2019).

For example, green and sustainable consumers might even consider engaging in activism (e.g., boycotts or bans; Friedman, 2017). For example, Coca-Cola's increased use of reusable packaging represents the company's reaction to consumer pressure to intensify its efforts to reduce plastic use by 2030 (Russ, d). Similarly, Dutch airline KLM had to address accusations of greenwashing, which led the company to discontinue its advertising campaign (Reuters, 2023a). Lufthansa and Air France were accused of engaging in similar business practices (Gruet, 2023). Similarly, the consumer brand Unilever claimed to have used overstated environmental claims to market its products, which were meant to mislead its consumers (Reuters, 2023b). Public attention was drawn to these events in all instances, supporting the media's role in raising awareness for corporate wrong-doings and shaping individual evaluations of selected incidents.

Conclusion

The present chapter outlined the media's central role in how individuals perceive the world. It is possible to draw individuals' attention to selected topics through agenda setting (i.e., *which issues* are relevant), thereby shaping the public's perceptions of important topics (McCombs & Venezuela, 2007). On the other hand, priming builds on the assumption that the media presents individuals with standards for forming their opinions. As such, it looks at *what* information is covered. Finally, framing suggests that the way information is presented is crucial in shaping individuals' perceptions of issues and topics, with a focus on *how* details are conveyed (Sonnett, 2019). In addition to directing public attention to issues, the media can also shape individuals' topic perceptions (McCombs & Venezuela, 2007).

Whereas these three theories reflect traditional mass media communication, which is not only one-directional but also does not allow for feedback and audience reactions, there is a need to adapt them to the more dynamic present-day media environment (Harder et al., 2017). For instance, instead of discussing (traditional) agenda setting, it might be more plausible to address two-way agenda setting, which has been discussed in several studies (Messner & Distaso, 2008; Russell Neuman et al., 2014). Moreover, while legacy media are still influential in dictating public news discourses, their print editions are not the only means to do so; instead, they increasingly rely on online news outlets (Ceron et al., 2016; Vonbun et al., 2016), which are based on interactivity. This change has even been found to influence the journalistic profession, which has, at least in parts, started to integrate audience feedback (Beckers & Harder, 2016).

Moreover, the influence that media audiences exercise must not be neglected (Weaver & Choi, 2017). Agenda setting is no longer a media task but involves the joint effort of multiple actors, including audience members. Hence, these dynamics must be considered in present-day media research, where consumers can contribute to raising topics on the public agenda as part of grassroots movements.

References

Al Mamun, A., Mohamad, M. R., Yaacob, M. R. B., & Mohiuddin, M. (2018). Intention and behavior toward green consumption among low-income households. *Journal of Environmental Management, 227*, 73–86. https://doi.org/10.1016/j.jenvman.2018.08.061

Aruguete, N., et al. (2007). The agenda-setting theory [La teoría agenda-setting.] *Scientific Information System, 1*(20), 44–50. https://doi.org/10.7764/cdi.20.111.

BBC. (2019, May 29). Ethical business: Young people behind a "rising trend." BBC News. Retrieved from https://www.bbc.com/news/uk-wales-48432432

Beckers, K., & Harder, R. (2016). Twitter just exploded": Social media as alternative vox pop. *Digital Journalism, 4*(7), 910–920. https://doi.org/10.1080/21670811.2016.1161493

Bilandzic, H., & Rössler, P. (2004). Life according to television. Implications of genre-specific cultivation effects: The gratification/cultivation model. *Communications, 29*(3), 295–326. https://doi.org/10.1515/comm.2004.020.

Biswas, A., & Roy, M. (2015). Leveraging factors for sustained green consumption behavior based on consumption value perceptions: Testing the structural model. *Journal of Cleaner Production, 95*, 332–340. https://doi.org/10.1016/j.jclepro.2015.02.042

Bruns, A. (2007). The future is user-led: The path toward widespread produsage. In *Proceedings of PerthDAC 2007: The 7th International Digital Arts and Culture Conference* (pp. 68–77). Curtin University of Technology.

Burmester, J., Sauermann, A., Spalek, K., & Wartenburger, I. (2018). Sensitivity to salience: Linguistic vs. visual cues affect sentence processing and pronoun resolution. *Language, Cognition and Neuroscience, 33*(6), 784–801. https://doi.org/10.1080/23273798.2018.1428758.

Carrington, M. J., Neville, B. A., & Whitwell, G. J. (2014). Lost in translation: Exploring the ethical consumer intention–behavior gap. *Journal of Business Research, 67*(1), 2759–2767. https://doi.org/10.1016/j.jbusres.2012.09.022

Carrington, M. J., Zwick, D., & Neville, B. (2016). The ideology of the ethical consumption gap. *Marketing Theory, 16*(1), 21–38. https://doi.org/10.1177/1470593115595674

Carrington, M., et al. (2021). Consumption ethics: A review and analysis of future directions for interdisciplinary research. *Journal of Bus Ethics, 168*, 215–238. https://doi.org/10.1007/s10551-020-04425-4

Carter, M. J. (2013). The hermeneutics of frames and framing: An examination of the media's construction of reality. *Sage Open, 3(2)*. https://doi.org/10.1177/2158244013487915.

Ceron, A., Curini, L., & Iacus, S. M. (2016). First-and second-level agenda setting in the Twittersphere: An application to the Italian political debate. *Journal of Information Technology & Politics, 13*(2), 159–174.

Chan, R. Y. (2001). Determinants of Chinese consumers' green purchase behavior. *Psychology & Marketing, 18*(4), 389–413. https://doi.org/10.1002/mar.1013

Chekima, B., Igau, A., Wafa, S. A. W. S. K., & Chekima, K. (2017). Narrowing the gap: Factors driving organic food consumption. *Journal of Cleaner Production, 166*, 1438–1447. https://doi.org/10.1016/j.jclepro.2017.08.086

Chen, M. F. (2020). Selecting environmental psychology theories to predict people's consumption intention of locally produced organic foods. *International Journal of Consumer Studies, 44*(5), 455–468. https://doi.org/10.1111/ijcs.12578

Cleveland, M., Kalamas, M., & Laroche, M. (2005). Shades of green: Linking environmental locus of control and pro-environmental behaviors. *Journal of Consumer Marketing, 22*(4), 198–212. https://doi.org/10.1108/07363760510605317

Close, C. (2021, May 18). *The global eco-wakening: How consumers are driving sustainability*. World Economic Forum. Retrieved from https://www.weforum.org/agenda/2021/05/eco-wakening-consumers-driving-sustainability/.

Cohen, B. C. (1963). *The press and foreign policy*. Princeton University Press.

Dikici, Z. Y., Çakrak, M., & Demirci, E. (2022). Green consumption values, social appreciation, and purchasing behavior. *Sustainability and Climate Change, 15*(3), 189–199. https://doi.org/10.1089/scc.2022.0003

Echegaray, F., & Hansstein, F. V. (2017). Assessing the intention-behavior gap in electronic waste recycling: The case of Brazil. *Journal of Cleaner Production, 142*, 180–190. https://doi.org/10.1016/j.jclepro.2016.05.064

Elkington, J. (1999). Cannibals with forks: The triple bottom line of 21st century business. *Oxford*. https://doi.org/10.1002/tqem.3310080106

Euronews. (2023, June 19). *Swiss citizens vote yes to "essential" new climate law*. Euronews. Retrieved from https://www.euronews.com/green/2023/06/19/electricity-sinkhole-or-important-step-forward-swiss-citizens-to-vote-on-new-climate-law.

Ewoldsen, D. R., & Rhodes, N. (2019). Media priming and accessibility. *Media Effects, 83–99*,. https://doi.org/10.4324/9780429491146-6

Fatemi, H., & Dube, L. (2021). The impact of corporate social responsibility (CSR) priming on consumers' consumption and lifestyle choices with intertemporal tradeoffs. *European Journal of Marketing, 55*(12), 3191–3220. https://doi.org/10.1108/EJM-04-2020-0250

Fischer, G. W., & Johnson, E. J. (1986). Behavioral decision theory and political decision making. *Political cognition: The 19th annual Carnegie Symposium on cognition* (pp. 55–65). Erlbaum.

Friedman, M. (2017). Using consumer boycotts to stimulate corporate policy changes: Marketplace, media, and moral considerations. In Monroe Friedman (Ed.) *Politics, products, and markets* (pp. 45–62). Routledge. https://doi.org/10.4324/9781315126937-4.

Gans, H. (1979). *Deciding what's news*. Pantheon.

Geiger-Oneto, S., & Arnould, E. J. (2011). Alternative Trade organization and subjective quality of life: The case of Latin American coffee producers. *Journal of Macromarketing, 31*(3), 276–290. https://doi.org/10.1177/0276146711405668

Goffman, E. (1974). *Frame analysis: An essay on the organization of experience*. Harvard University Press.

Gruet, S. (2023, December 12). Dove and marmite maker unilever faces investigation over green claims. www.bbc.com. Retrieved from https://www.bbc.com/news/business-67658043.

Gummerus, J., Liljander, V., & Sihlman, R. (2017). Do ethical social media communities pay off? An exploratory study of the ability of facebook ethical communities to strengthen consumers' ethical consumption behavior. *Journal of Business Ethics, 144*, 449–465. https://doi.org/10.1007/s10551-015-2830-y

Halder, P., Hansen, E. N., Kangas, J., & Laukkanen, T. (2020). How national culture and ethics matter in consumers' green consumption values. *Journal of Cleaner Production, 265*, 121754. https://doi.org/10.1016/j.jclepro.2020.121754

Hanson, R. E. (2016). *Mass communication: Living in a media world*. Sage publications.
Harder, R. A., Sevenans, J., & Van Aelst, P. (2017). Intermedia agenda setting in the social media age: How traditional players dominate the news agenda in election times. *The International Journal of Press/politics, 22*(3), 275–293. https://doi.org/10.1177/1940161217704969
Hassan, S. M., Rahman, Z., & Paul, J. (2022). Consumer ethics: A review and research agenda. *Psychology & Marketing, 39*(1), 111–130.
Haws, K. L., Winterich, K. P., & Naylor, R. W. (2014). Seeing the world through green-tinted glasses: Green consumption values and responses to environmentally friendly products. *Journal of Consumer Psychology, 24*(3), 336–354.
Higgins, E. T., Bargh, J. A., & Lombardi, W. J. (1985). Nature of priming effects on categorization. *Journal of Experimental Psychology: Learning, Memory, & Cognition, 11*, 59–69. https://doi.org/10.1037/0278-7393.11.1.59
Hoewe, J. (2020). Toward a theory of media priming. *Annals of the International Communication Association, 44*(4), 312–321. https://doi.org/10.1080/23808985.2020.1815232
Homar, A. R., & Cvelbar, L. K. (2021). The effects of framing on environmental decisions: A systematic literature review. *Ecological Economics, 183*, 106950. https://doi.org/10.1016/j.ecolecon.2021.106950
IBM (2022). IBM report: More consumers driven by purpose than by value. [online] Retrieved from https://businesschief.eu/leadership-and-strategy/ibm-report-more-consumers-driven-purpose-value.
Iyengar, S. & Kinder, D.R. (1987). *News that matters: Television and American opinion*. University of Chicago Press.
Joshi, Y., & Rahman, Z. (2015). Factors affecting green purchase behavior and future research directions. *International Strategic Management Review, 3*(1–2), 128–143. https://doi.org/10.1016/j.ism.2015.04.001
Lillo-Viedma, F., Severino-González, P., Santander-Ramírez, V., García, L. Y., Guiñez-Cabrera, N., & Astorga-Bustos, N. (2022). Corporate social responsibility and social network analysis: Unionized workers' perceptions. *Sustainability, 14*(7), 4320. https://doi.org/10.3390/su14074320
Lu, L. C., Chang, H. H., & Chang, A. (2015). Consumer personality and green buying intention: The mediate role of consumer ethical beliefs. *Journal of Business Ethics, 127*, 205–219. https://doi.org/10.1007/s10551-013-2024-4
Luhmann, N. (1995). Die Realität der Massenmedien. In: Die Realität der Massenmedien. Nordrhein-Westfälische Akademie der Wissenschaften, Vol 333. VS Verlag für Sozialwissenschaften, Wiesbaden. https://doi.org/10.1007/978-3-663-16287-2_1.

Mazza, R., Zavarrone, E., Olivieri, M., & Corsaro, D. (2022). A text mining approach for CSR communication: An explorative analysis of energy firms on Twitter in the postpandemic era. *Italian Journal of Marketing, 2022*(3), 317–340. https://doi.org/10.1007/s43039-022-00050-3

McCombs, M. (1997). New frontiers in agenda setting: Agendas of attributes and frames. *Mass Communication Review, 24*(1/2), 32–52.

McCombs, M. (2005). A look at agenda-setting: Past, present and future. *Journalism Studies, 6*(4), 543–557. https://doi.org/10.1080/14616700500250438

McCombs, M. E., & Shaw, D. L. (1972). The agenda-setting function of mass media. *Public Opinion Quarterly, 36*(2), 176–187. https://doi.org/10.1086/267990

McCombs, M., Llamas, J. P., Lopez-Escobar, E., & Rey, F. (1997). Candidate images in Spanish elections: Second-level agenda-setting effects. *Journalism & Mass Communication Quarterly, 74*(4), 703–717. https://doi.org/10.1177/107769909707400404

McCombs, M. E., Shaw, D. L., & Weaver, D. H. (2014). New directions in agenda-setting theory and research. *Mass Communication and Society, 17*, 781–802. https://doi.org/10.1080/15205436.2014.964871

McCombs, M., & Valenzuela, S. (2020). *Setting the agenda: Mass media and public opinion*. John Wiley & Sons.

Messner, M., & Distaso, M. W. (2008). The source cycle: How traditional media and weblogs use each other as sources. *Journalism Studies, 9*(3), 447–463. https://doi.org/10.1080/14616700801999287

Mongabay. (2023, July 27). *The circular economy: Sustainable solutions to solve planetary overshoot?* Mongabay Environmental News. Retrieved from https://news.mongabay.com/2023/07/the-circular-economy-sustainable-solutions-to-solve-planetary-overshoot/.

Moy, P., Tewksbury, D., & Rinke, E. M. (2016). Agenda-setting, priming, and framing. *The International Encyclopedia of Communication Theory and Philosophy, 1–13*,. https://doi.org/10.1002/9781118766804.wbiect266

Muncy, J. A., & Vitell, S. J. (1992). Consumer ethics: An investigation of the ethical beliefs of the final consumer. *Journal of Business Research, 24*(4), 297–311. https://doi.org/10.1016/0148-2963(92)90036-B

Mutum, D. S., Ghazali, E. M., & Wei-Pin, W. (2021). Parallel mediation effect of consumption values and the moderation effect of innovativeness, in predicting the influence of identity on green purchasing behavior. *Journal of Consumer Behavior, 20*(3), 827–844. https://doi.org/10.1002/cb.1913

Neuman, W. R., Guggenheim, L., Jang, S. M., & Bae, S. (2014). Theorizing the dynamics of public attention: Agenda setting theory meets big data. *Journal of Communication, 64*(2), 193–214. https://doi.org/10.1111/jcom.12088

Nguyen, T. N., Lobo, A., & Nguyen, B. K. (2017). Young consumers' green purchase behavior in an emerging market. *Journal of Strategic Marketing, 26*(7), 583–600. https://doi.org/10.1080/0965254X.2017.1318946

Nowak, E. (2016). *Agenda-setting theory and the new media*. Retrieved from https://archiwum.studiamedioznawcze.eu/Numery/2016_3_66/nowak-en.pdf. https://doi.org/10.33077/uw.24511617.ms.2016.66.449.

Pan, Z., & Kosicki, G. M. (1993). Framing analysis: An approach to news discourse. *Political Communication, 10*, 55–75. https://doi.org/10.1080/10584609.1993.9962963

Paus-Hasebrink, I., Kulterer, J., Sinner, P., Paus-Hasebrink, I., Kulterer, J., & Sinner, P. (2019). The role of media within young people's socialization: A theoretical approach. *Social Inequality, Childhood and the Media: A Longitudinal Study of the Mediatization of Socialization, 45–75*. https://doi.org/10.1007/978-3-030-02653-0_3

Price, V., & Tewksbury, D. (1997). News values and public opinion: A theoretical account of media priming and framing. *Progress in communication sciences*, 173–212.

Reuters. (2023a, December 12). UK to probe Unilever's environmental claims as part of greenwashing crackdown. Reuters. Retrieved from https://www.reuters.com/world/uk/britain-scrutinise-unilevers-green-claims-household-items-2023-12-12/

Reuters. (2023b, December 19). *KLM defends advertisements in Dutch court, denies "greenwashing."* Reuters. Retrieved from https://www.reuters.com/business/aerospace-defense/klm-dutch-court-answer-greenwashing-allegation-2023-12-19/.

Roskos-Ewoldsen, D. R., Klinger, M., & Roskos-Ewoldsen, B. B. (2007). Media priming. In R. W. Preiss, B. M. Gayle, N. Burrell, M. Allen, & J. Bryant (Eds.), *Mass media theories and processes: Advances through meta-analysis* (pp. 53–80). Lawrence Erlbaum Associates.

Roskos-Ewoldsen, D. R., Roskos-Ewoldsen, B. B., & Carpentier, F. D. (2009). Media priming: An updated synthesis. In J. Bryant & M. B. Oliver (Eds.), *Media effects: Advances in theory and research* (3rd ed., pp. 74–93). Routledge.

Russ, H. (2022, February 15). Coca-Cola, criticized for plastic pollution, pledges 25% reusable packaging. *Reuters*. Retrieved from https://www.reuters.com/business/sustainable-business/coca-cola-criticized-plastic-pollution-pledges-25-reusable-packaging-2022-02-10/.

Scheufele, D. A. (1999). Framing as a theory of media effects. *Journal of Communication, 49*(1), 103–122. https://doi.org/10.1111/j.1460-2466.1999.tb02784.x

Scheufele, D. A. (2000). Agenda-setting, priming, and framing revisited: Another look at cognitive effects of political communication. *Mass Communication &*

Society, 3(2–3), 297–316. https://doi.org/10.1207/S15327825MCS0323_07
Scheufele, D. A., & Tewksbury, D. (2007). Framing, agenda setting, and priming: The evolution of three media effects models. *Journal of Communication, 57*(1), 9–20. https://doi.org/10.1111/j.0021-9916.2007.00326.x
Schulz, W. (2003). Mediennutzung und Umweltbewusstsein: Dependenz-und Priming-Effekte. *Publizistik, 4*(48), 387–413. https://doi.org/10.1007/s11616-003-0108-3
Shoemaker, P.J., & Reese, S.D. (1995). *Mediating the message: Theories of influences on mass media content.* Longman Publications.
Sivapalan, A., von der Heidt, T., Scherrer, P., & Sorwar, G. (2021). A consumer values-based approach to enhancing green consumption. *Sustainable Production and Consumption, 28*, 699–715. https://doi.org/10.1016/j.spc.2021.06.013
Skaidrė, Ž. (2019). Grassroots activism and sustainable development. In: Leal Filho, W. (Ed.) Encyclopedia of Sustainability in Higher Education. Springer, Cham. https://doi.org/10.1007/978-3-319-63951-2_281-1.
Sonnett, J. (2019). Priming and framing: Dimensions of communication and cognition. In Wayne H. Brekhus & Gabe Ignatow (Eds.), *The Oxford Handbook of Cognitive Sociology*, Oxford Handbooks (2019; online ed. Oxford Academic, July 2 2019), https://doi.org/10.1093/oxfordhb/9780190273385.013.13.
Tanford, S., Kim, M., & Kim, E. J. (2020). Priming social media and framing cause-related marketing to promote sustainable hotel choice. *Journal of Sustainable Tourism, 28*(11), 1762–1781. https://doi.org/10.1080/09669582.2020.1760287
Taylor, M., Watts, J., & Bartlett, J. (2019, September 27). *Climate crisis: 6 million people join latest wave of global protests.* The Guardian. Retrieved from https://www.theguardian.com/environment/2019/sep/27/climate-crisis-6-million-people-join-latest-wave-of-worldwide-protests.
Uddin, S. F., & Khan, M. N. (2016). Exploring green purchasing behavior of young urban consumers: Empirical evidence from India. *South Asian Journal of Global Business Research, 5*(1), 85–103. https://doi.org/10.1108/SAJGBR-12-2014-0083
Van Doorn, J., & Verhoef, P. C. (2015). Drivers of and barriers to organic purchase behavior. *Journal of Retailing, 91*(3), 436–450. https://doi.org/10.1016/j.jretai.2015.02.003
Vitell, S. J. (2015). A case for *consumer* social responsibility (CnSR): Including a selected review of consumer ethics/social responsibility research. *Journal of Bussiness Ethics, 130*, 767–774. https://doi.org/10.1007/s10551-014-2110-2

Vonbun, R., Königslöw, K. K. V., & Schoenbach, K. (2016). Intermedia agenda-setting in a multimedia news environment. *Journalism, 17*(8), 1054–1073. https://doi.org/10.1177/1464884915595475

Wang, P., & Kuah, A. T. (2018). Green marketing cradle-to-cradle: Remanufactured products in Asian markets. *Thunderbird International Business Review, 60*(5), 783–795. https://doi.org/10.1002/tie.21925

Wanta, W., & Tarasevich, S. (2019). Agenda-setting and priming theories. *The International Encyclopedia of Journalism Studies,* 1–6. 10.1002/9781118841570.iejs0014.

Weaver, D. H., & Choi, J. (2017). The media agenda. Who (or What) Sets It? *The Oxford Handbook of Political Communication*, 359.

Weder, F. (2012). *Die CSR-Debatte in den Printmedien. Anlässe, Themen, Deutungen*. Wien: Facultas WU Wien.

Weder, F. (2017). CSR as common sense issue? A theoretical exploration of public discourses, common sense and framing of corporate social responsibility. In: Diehl, S., Karmasin, M., Mueller, B., Terlutter, R., & Weder, F. (Eds.) *Handbook of Integrated CSR Communication*. CSR, Sustainability, Ethics & Governance. Springer, Cham. https://doi.org/10.1007/978-3-319-44700-1_2.

Weimann, G., Weiss-Blatt, N., Mengistu, G., Tregerman, M. M., & Oren, R. (2014). Reevaluating "the end of mass communication?" *Mass Communication and Society, 17*(6), 803–829. https://doi.org/10.1080/15205436.2013.851700

Weimann, G., & Brosius, H. B. (2015). A new agenda for agenda-setting research in the digital era. In Gerhard Vowe & Philipp Henn (Eds.) *Political Communication in the Online World* (pp. 26–44). Routledge. https://doi.org/10.4324/9781315707495-3.

White, K., Hardisty, D., & Habib, R. (2019). *The elusive green consumer*. Harvard Business Review. Retrieved from https://hbr.org/2019/07/the-elusive-green-consumer.

Wirt, F. M., Micheletti, M., Follesdal, A., & Stolle, D. (2017). Politics, products and markets exploring political consumerism past and present. *Routledge*. https://doi.org/10.4324/9781315126937

Yan, L., Keh, H. T., & Wang, X. (2021). Powering sustainable consumption: The roles of green consumption values and power distance belief. *Journal of Business Ethics, 169*, 499–516. https://doi.org/10.1007/s10551-019-04295-5

Zeng, T., Audrain-Pontevia, A. F., & Durif, F. (2021). Does corporate social responsibility affect consumer boycotts? A cost–benefit approach. *Corporate Social Responsibility and Environmental Management, 28*(2), 796–807. https://doi.org/10.1002/csr.2089

CHAPTER 12

Ethical Determinants of Consumer Behavior and Responsible Marketing Practice

Danijela Križman and Darko Rendulić

Abstract This chapter focuses on 'ethical living' as one of the eight megatrends (Boumphrey & Brehmere, 2017). Consumers and business decisions are increasingly guided by concerns about the environment, sustainability, animal welfare, production, labor practices, and a desire to impact communities and people positively. Researchers posit that *"with the same ostensible value of quality and price, customers are likely to prefer the brand of the socially responsible organization"* (Mandal and Banerjee, 2018). The contemporary marketing literature suggests that *"the responsible behavior of any company is an integral part of the marketing process and is generally termed responsible marketing"* (Bhattacharya et al., 2007). Because of *"acceptance to both internal and external stakeholders, responsible marketing, unlike traditional marketing, may serve as a better means*

D. Križman (✉)
Faculty of Economics and Tourism "Dr Mijo Mirković", Juraj Dobrila University of Pula, Pula, Croatia
e-mail: dkrizman@unipu.hr

D. Rendulić
Karlovac University of Applied Sciences, Karlovac, Croatia

I. Pavić et al. (eds.), *Bioethics and Social Ethics in The Modern World*,
https://doi.org/10.1007/978-3-031-86418-6_12

of communication for the companies" (Mandal and Banerjee, 2018), determinants of responsible marketing as a business concept are discussed. Additionally, selected responsible marketing practices are identified and analyzed. Recommendations for future research and marketing managers are formulated and presented.

Keywords Megatrends · Ethical living · LOHAS consumers · Business concept · Responsible marketing

Introduction

In the contemporary business landscape, characterized by rapid change and intensified competition, recognizing and adapting to emerging megatrends is critical for industries seeking sustained economic growth, job creation, and ongoing national development. Megatrends represent significant shifts in consumer behavior with global and cross-industry implications. According to Euromonitor International, one of the eight megatrends that will significantly impact industries and consumers by 2030 is "ethical living" (Boumphrey & Brehmere, 2017). This trend indicates that consumer and business decisions are increasingly driven by ethical considerations, including environmental sustainability, animal welfare, responsible production and labor practices, and a desire to impact communities positively.

With the rise of the "ethical living" megatrend, a new consumer demographic, lifestyle of health and sustainability (LOHAS), has emerged in the market. This term, first coined by Ray and Anderson (2001), describes a group of consumers who are environmentally conscious, socially attuned, and mindful of the broader impact of their choices at the personal, community, and planetary levels (Pícha and Navrátil 2019). LOHAS consumers prioritize companies that demonstrate responsible behavior, seeking transparency, and ethical practices in their business dealings.

The contemporary marketing literature underscores the importance of responsible marketing, with Bhattacharya et al. (2007) suggesting that a company's responsible behavior is an integral component of the marketing process. Unlike traditional marketing, responsible marketing emphasizes transparency, fairness, and accountability, from product development and communication to the broader impact of marketing activities (Mandal & Banerjee, 2018).

This research aims to achieve the following objectives: (1) to identify ethical determinants of consumer behaviour; (2) to determine the key characteristics and unique attributes of the LOHAS consumer segment, along with an estimation of its market size; (3) to describe responsible marketing as a new business concept; and (4) to identify and analyze best practices in responsible marketing. The findings from this study are intended to offer insights for future research and provide actionable recommendations for marketing managers and other stakeholders in the business community.

Ethical Determinants of Consumer Behavior and the LOHAS Market

The term "ethical consumer" gained popularity in 1989 owing to the Ethical Consumer Magazine.[1] Since then, it has become a widely used generic term. An ethical consumer involves selecting products and services sourced, produced, and distributed that adhere to ethical standards. Consumer ethics refers to the moral principles guiding a consumer's buying choices and actions.

As a research domain, consumer ethics has experienced considerable growth since 2000 (Hassan et al., 2022). Interest in consumer ethics grew as researchers and practitioners realized that (1) consumers play an integral role in the seller–buyer relationship and (2) fostering a genuinely ethical marketplace is essential for extending marketing transactions beyond short-term, one-time exchanges (Hennig-Thurau et al., 2002; Rao & Al-Wugayan, 2005). A deeper understanding of consumer ethics is important for practitioners because it can help develop strategies to reduce unethical consumer behavior while promoting ethical awareness and responsibility (Arli et al., 2018; Chatzidakis & Maclaran, 2020).

Numerous factors influence consumer ethics. Hassan et al. (2022) categorized them into antecedents, mediators, moderators, and output variables. There are nine distinct groups of *antecedent* variables influencing consumers' ethical behavior (Hassan et al., 2022):

[1] See: https://www.ethicalconsumer.org/about-us (2024).

- Personal factors, i.e., religiosity, attitudes (attitudes toward business or digital piracy), materialism, moral philosophies, spirituality, Machiavellianism, etc.;
- Social and interpersonal factors include social norms and perceived behavioral control, moral intensity, social attachment, parents' religiosity, parenting style, peer support, and others;
- Emotions, e.g., guilt, pride, fear, and anger;
- Intuitive factors, e.g., intuition and implicit attitudes;
- Firm-related factors, such as corporate associates, CSR, product attributes, ethical claims, firm size, and payment timing;
- Cultural factors, for example, collectivism;
- Coping factors such as neutralization techniques;
- Deterrence factors, including factors such as fear of legal consequences and perceived risk;
- Other factors, such as values, induced hypocrisy, and impediment.

Other *"socio-demographic (e.g., migration background) and psychographic variables (e.g., individualism) should also be considered antecedents that influence consumers' ethical judgment"* (Hassan et al., 2022: 13). Research should study *"the influence of moral attentiveness, envy, generosity and compassion, self-accountability, self concept, ethical blindness, and self-esteem as antecedents of consumer ethical behavior"* (Hassan et al., 2022: 13).

Some variables *mediate* the relationship between the antecedents and outcomes of consumer ethics. Hassan et al. (2022) grouped those variables into seven clusters:

- Personal factors include moral philosophy, materialism, ethnocentricity, social goodwill, Machiavellianism, monetary factors, and attitudes toward the environment.
- Emotions include anticipated guilt, empathy, and happiness.
- Firm-related factors include relationship quality and outcome, trust in ethical advertising, and trust in product information.
- Social and interpersonal factors include benefits from the Facebook community, susceptibility, and normative influence.
- Deterrence factors that include perceived risk.
- Coping factors
- Other factors, such as perceived justice and implementation plans.

A literature review revealed that *"there is a need to study the mediating role of various emotional (e.g., guilt, empathy), social and interpersonal (e.g., social attachment, social identity), firm-related (e.g., customer–firm relationship), cultural (e.g., long-term orientation) and coping (e.g., neutralization techniques) factors to explain the processes through which various antecedents impact the ethical beliefs, intentions and behavior of consumers"* (Hassan et al., 2022: 13).

Certain variables can *moderate* consumers' ethical behavior. Hassan et al. (2022) classified those variables into the following six categories:

- Personal factors, i.e., construal level, ethnocentricity, and ability;
- Social factors and interpersonal factors, for example, the role of the third-person perspective and attention to social comparison information;
- Firm-related factors include self-benefit appeals in advertising, postponed payments, etc.
- Cultural factors, i.e., power distance, uncertainty avoidance, and Hofstede's other cultural dimensions;
- Coping factors, such as neutralization strategies;
- Other factors include the potential for justice restoration and the purchase situation.

Factors such as peer pressure, social comparison, and demographic variables, including age and gender, can also be analyzed as moderators in the context of unethical consumer behavior (Chowdhury, 2019; Viglia et al., 2019).

The *outcome* variables examined in consumer ethics research are broadly classified as follows (Hassan et al., 2022):

- Beliefs—consumer beliefs toward unethical actions and ethical and prosocial actions.
- Intentions—positive/ethical intentions (i.e., intentions to recycle, intentions to buy products that make ethical claims, intentions to buy energy-efficient portals, willingness to choose ethical products"), and harmful or unethical intentions (such as piracy intentions);
- Behavior—ethical behavior and unethical behavior.

While practitioners have a range of predictors to analyze a particular market, lifestyle variables have only recently started to be used to create market segments (Pícha & Navratil, 2019). The lifestyle of health and sustainability (LOHAS) is currently one of the most recognized consumer lifestyles, and the key value categories defining this lifestyle are as follows (Lendvai et al., 2022):

- authentic values,
- environmentally conscious values,
- individual values,
- ethical values,
- health-conscious values,
- emotional values and
- spiritual values.

Members of the LOHAS market pay close attention to healthy goods produced ethically (Yeh & Chen, 2011). Francis and Hoefel (2018) reported that 70% of this group only purchase goods from companies they deem ethical, and 80% never purchase anything from a company that has demonstrated contradictory behavior or has been involved in a scandal.

According to Peyron (2010), approximately 100 million people worldwide follow a LOHAS lifestyle, and approximately 20% of the population in Europe may be labeled LOHAS consumers. In developed countries, the proportion of consumers prioritizing sustainability in their purchasing decisions may reach 25% of the population (Szakály et al., 2015). Therefore, to become the brand of choice for the LOHAS market, companies need not only to communicate but also to act with ethical standards.

Today, we are witnessing the emergence of a new consumer generation, Gen Z (colloquially referred to as Zoomers, born between 1997 and 2012), representing approximately 32% of the global population, making it the most significant consumer generation to date. Companies must understand their needs to foster growth as this generation matures and makes significant purchasing decisions for their families and futures. Reportedly, 70% of Gen Z say that they try to purchase products from companies they consider ethical (Francis & Hoefel, 2018). By analyzing comprehensive data on Gen Z's preferences, desires, and demands, companies can make informed decisions to earn the trust and loyalty of this emerging group of consumers, meeting their expectations for ethical consumption.

Responsible marketing as a business concept

Many authors emphasize that consumers are saturated with enormous amounts of data that cannot be processed adequately to make "correct" buying decisions. The marketers are to blame for that, but they only follow subordinate top management strategies. They need a universal roadmap of what creates a better world through socially responsible marketing means. Missing that, lawful maximizing profit remains pivotal "raison d'être." Laczniak and Shulz (2020) used macro-level analysis and normative ethical theory to delineate and explicate a socially responsible marketing doctrine (SRM). This concept offers three essential elements of SRM: corporate citizenship, stakeholder orientation, and social/ecological sustainability, which could be explained as follows:

SRM (socially responsible marketing) = CC (corporate citizenship) + SO (stakeholder orientation) + SES (social/ecological sustainability).

The CC is exposed to a conflict among legislatures when a corporation operates, and the SO bears a conflict of interest with the stakeholders' groups. Moreover, SESs require a holistic approach and are primarily contained in the UN's SDGs, particularly *SDG 12, "Responsible Consumption and Production,"* which can be applied globally. Mayasari et al. (2023) emphasize the legal frame as a criterion for implementing marketing strategies and claim the obligation of marketers to give customers their rights. Their study suggested that moral judgment and the ethics code are the determinant factors of ethical behavior.

Best Responsible Marketing Practices

To achieve results promised to stakeholders, companies apply published company social responsibility (CSR) programs that face numerous obstacles. Bhattacharya et al. (2007) emphasize the human resource problem of finding, educating, and retaining efficient personnel as an important communication channel with buyers: *"In general, CSR initiatives, particularly those that are specific (and that we primarily focus on in our research), can be effective in building and sustaining a talented employee base because they reveal the values of a company and thus can be part of the 'Employee Value Proposition' or the holistic sum of everything people experience and receive while they are a part of a company."*

They indicated four practices which prevent effective internal marketing leverage concerning CSR:

- Many organizations withhold comprehensive details about their CSR initiatives from their employees, failing to communicate the scope and specifics of these efforts consistently and transparently.
- Companies often design their CSR strategies without explicitly addressing the varied needs of their workforce that such programs could potentially fulfill.
- Firms frequently lack insight into the psychological processes through which their CSR efforts translate into beneficial employee outcomes—such as enhanced commitment, increased productivity, and reduced turnover.
- Organizations adopt a rigid, top-down approach when formulating, implementing, and sustaining their CSR programs, often compelling employees to participate rather than inviting them to engage on their terms.

This implies that companies have not recognized or used the potential of CSR to optimize internal and external communications. In particular, barriers to understanding their core values and ethics are removed. Top management's subordinating rules for achieving results are far from direct contact with market actors. One way to ensure supremacy is to control the situation, but how to survive morally correctly in free competition requires a deeper understanding of the market. Schlipzand and de Jong (2023) provided a short overview of the theoretical attitudes. Since moral behavior is context-dependent, large-scale studies have provided results that had to be observed from different optics. Hence, Schlipzand and de Jong (2023) focused on civic morality, which they considered crucial for a well-functioning market society. Their results indicated a positive relationship between an increase in market society and civic morality despite the area investigated. A related standard indicator is the economic freedom index (EFI). Schlipzand and de Jong (2023) summarized these findings, showing that economic freedom increases trust tolerance and reduces antisemitism. Positive scores for market economies correlate with prosocial behavior (Choi & Storr, 2019), such as altruism and voluntary work.

Griffith (2017) contemplates the potential of behavioral economics for use in marketing. She accepts cultural differences as differentiators when evaluating the market's moral aspects. What is considered manipulation in one context is being used as a legitimate policy tool in others. The general consensus is that behavioral economics is a tool at the disposal of policymakers and businesspeople to influence behavior.

Akhtar (2024) noted that consumer preferences shifted toward sustainability and ethics, considering those as new challenges and opportunities. These preferences reshape industries, driving them toward eco-friendly products and ethical practices. This is explained through seven aspects:

1. Changing consumer landscape—According to a survey by Nielsen, 73% of global consumers say that they would change their consumption habits to reduce their environmental impact.
2. Business imperatives—addressing sustainability and ethics in operations to meet consumer expectations and remain competitive—research by McKinsey & Company indicates that companies with substantial ESG performance outperform their peers financially.
3. Market opportunities—the shift toward sustainable and ethical consumer behavior is creating new market opportunities for businesses
4. Supply chain transparency—demand for greater transparency and accountability from businesses regarding their supply chains and manufacturing processes
5. Brand loyalty and reputation—embracing sustainability and ethics can enhance brand loyalty and reputation, fostering consumer trust and loyalty.
6. Regulatory pressures—regulations and incentives to promote sustainability and ethical practices in business operations
7. In the future, sustainable and ethical consumer behavior continues to gain momentum; businesses must adapt and innovate to meet evolving consumer expectations. By embracing sustainability and ethics, companies can drive positive change, build resilient business models, and create a more sustainable and equitable future for all.

Marketers are trying to find ways to communicate the performance of goods or services in terms of spending less energy and raw materials, reducing waste by using biodegradable materials, decreasing CO_x, NO_x, and SO_x emissions, etc. Kovalenko (2024) presents additional examples of ethical marketing strategies:

1. Allbirds: The company supports forestation, advocates sustainable agricultural practices, and builds an innovative supply chain using renewable materials, green energy, and regenerative agriculture.
2. Patagonia: *"Do not buy this jacket."* The brand's philosophy is about being green, with ethical marketing being a natural part of its strategy.
3. Warby Parker: *"Buy a pair, give a pair."* By reducing prescribed eyeglass prices, the company successfully increases annual revenue by 38%.
4. TOMS Shoes: *"One Day Without Shoes"*—a creative campaign that provoked social responsibility and solidarity. Anyone could participate by posting a photo of their bare feet on Instagram with the hashtag #WithoutShoes. One photo equaled the donation of a pair of shoes to a child in need.
5. Faguo: Engaging in a generation against global warming. As clothing companies, they claim to maintain eco-friendly standards as follows:
   - offsetting its carbon footprint,
   - reducing harmful emissions from production,
   - a tree planted for every item sold (2 + million thus far),
   - promoting reasonable consumerism by repairing products,
   - collecting and recycling second-hand pieces from customers.
6. On the basis of an Instagram campaign, Pela is the world's first biodegradable and compostable phone case.
7. Asphalte: Zero overproduction with a preorder model—no stocks, activates upon the order
8. Yes Straws: Providing an alternative to plastic straws by producing biodegradable straws made from recycled materials.
9. Cusa Tea & Coffee:
   - innovative recycling program for customers,
   - support of environmental projects,
   - use of eco-friendly packaging materials,
   - the used tea and coffee sticks were exchanged for coupons.

10. Everlane: "Radical transparency" in the value chain.
11. Illy Coffee: *"Reforestation, farmer education, sustainability—a leader in advancing ethical business standards,"* which applies to both operations and marketing by funding rural schools and nursing plant species and providing water treatment systems for coffee producers in Nicaragua and Honduras

Conclusion

By providing the relevant, transparent information needed for informed decision-making, adaptable marketing practices are vital in bridging the knowledge gap between manufacturers/suppliers and buyers/consumers. However, market-driven environments often encourage self-interest while diminishing moral obligations (Schlipzand & de Jong, 2023). At the same time, ethics evolves as we continually evaluate the rational basis for our moral judgments. Consequently, it shapes consumer choices and broader societal interactions, which are reinforced by the influence of various media outlets. In situations where consumers have limited access to education, marketers and influencers bear an even greater responsibility to promote ethical awareness.

Ethical living, which includes the conscientious consumption of ethically produced goods, presents an additional challenge for marketers—one that goes beyond merely outpacing competitors (Boumphrey & Brehmere, 2017). By recognizing and anticipating market megatrends, marketers can respond effectively to shifting consumer demands. Ultimately, building consumer trust and fostering responsible practices depend on implementing publicly accessible marketing instruments, adhering to robust codes of ethics, and enforcing clear quality assurance policies.

References

Akhtar, T. (2024). The impact of sustainable and ethical consumer behavior on businesses. Retrieved from https://www.linkedin.com/pulse/impact-sustainable-ethical-consumer-behavior-akhtar-cem--1w5pf/?trk=articles_directory.

Arli, D., Tan, L. P., Tjiponto, F., & Yang, L. (2019). Exploring consumers' purchase intention toward green products in an emerging market: The role of consumers' perceived readiness. *International Journal of Consumer Studies, 42*(4). https://doi.org/10.1111/ijcs.12432.

Bhattacharya, C. B., Sen, S., & Korschun, D. (2007). Corporate social responsibility as an internal marketing strategy. *Sloan Management Review, 49*(1), 1–29.

Boumphrey, S., & Brehmere, Z. (2017). *Megatrend analysis: Putting the consumer at the heart of business*. Euromonitor International.

Chatzidakis, A., & Maclaran, P. (2020). Gendering consumer ethics. *International Journal of Consumer Studies*. https://doi.org/10.1111/ijcs.12567

Choi, V. H., & Storr, G. S. (2019). *Do Markets Corrupt Our Morals?*. Springer International Publishing.

Chowdhury, R. M. (2019). The moral foundations of consumer ethics. *Journal of Business Ethics, 158*(3), 585–601. https://doi.org/10.1007/s10551-017-3676-2

Francis, T., & Hoefel, F. (2018), True Gen': Generation Z and its implications for companies. *Consumer Packaged Goods,* McKinsey&Co.

Griffith, L. (2017). Behavioral economics in marketing: Is it ethical? Retrieved from https://www.linkedin.com/pulse/behavioural-economics-marketing-ethical-liz-griffith/.

Hassan, S. M., Rahman, Z., & Paul, J. (2022). Consumer ethics: A review and research agenda. *Psychology & Marketing 39*(1), 111–130. https://doi.org/10.1002/mar.21580

Hennig-Thurau, T., Gwinner, K. P., & Gremler, D. D. (2002). Understanding relationship marketing outcomes: An integration of relational benefits and relationship quality. *Journal of Service Research, 4*(3), 230–247. https://doi.org/10.1177/1094670502004003006

Kovalenko, Oleksii. (2024). *Ethical marketing done right: 11 examples of brands thinking differently.* Retrieved from https://wisepops.com/blog/ethical-marketing.

Krupka, Z., & Kelecic, D. (2011). Istraživanje stavova LOHAS segmenta prema zelenim markama. *Tržiste, XXIII*(2), 247-262.

Laczniak, G., & Shultz, C. (2020). Toward a doctrine of socially responsible marketing (SRM): A macro and normative-ethical perspective. *Journal of Macromarketing, 41*(2), 201–231. https://doi.org/10.1177/0276146720963682

Lendvai, M. B., Kovács, I., Balázs, B. F., & Beke, J. (2022). Health and environment conscious consumer attitudes: Generation Z segment personas according to the LOHAS model. *Social Sciences, 11*, 269. https://doi.org/10.3390/socsci11070269

Mandal, K., & Banerjee, S. (2018). Responsible marketing and its impact on business performance: A longitudinal study. *Journal of Nonprofit & Public Sector Marketing, 115–138*. https://doi.org/10.1080/10495142.2018.1526736

Mayasari, L., Wulandari, D., Maharani, A., & Wiadi, I. (2013). Determinants of ethical behavior: An empirical study of implementing marketing strategies. *Development Research Management, 8*(1), 2–19.

Nefat, A. (2019). *Strategies of sustainable marketing [Strategije održivog marketinga]*. Universitiy of Juraj Dobrila in Pula.

Pícha, K., & Navratil, J. (2019). The factors of Lifestyle of Health and Sustainability influencing pro-environmental buying behaviour. *Journal of Cleaner Production, 234*, 233–241. https://doi.org/10.1016/j.jclepro.2019.06.072

Rao, C. P., & Al-Wugayan, A. A. (2005). Gender and cultural differences in consumer ethics in a consumer-retailer interaction context. *Journal of International Consumer Marketing, 18*(1–2), 45–71. https://doi.org/10.1300/J046v18n01_03

Schilpzand, A., & de Jong, E. (2023). Do market societies undermine civic morality? An empirical investigation into market societies and civic morality across the globe. *Journal of Economic Behavior and Organization., 208*, 39–60. https://doi.org/10.1016/j.jebo.2023.01.020

Viglia, G., Tassiello, V., Gordon-Wilson, S., & Grazzini, L. (2019). Predicting consumers' cheating behavir. The role of mental representation of goods and psychological ownership. *Psychology and Marketing, 36*(11), 1039–1045. https://doi.org/10.1002/mar.21254.

Yeh, N. C., & Chen, Y. J. (2011). On the everyday life information behavior of LOHAS consumers: A perspective of lifestyle. *Journal of Educational Media and Library Science, 48*(4), 489–510.

CHAPTER 13

Food Consumption Patterns and Sustainability: Results of Empirical Research

Mirna Leko Šimić, Marija Čutura, and Marina Stanić

Abstract Sustainable behavior refers to an awareness of the long-term consequences of individuals' everyday actions on the natural and/or social environment. This study focuses on sustainable food consumption, which refers to eating habits, choices, and attitudes that benefit the environment, health, and society. This study aimed to gain deeper insight into food consumption patterns from a sustainability perspective and to relate these patterns to knowledge (or lack of knowledge) about sustainable

M. L. Šimić (✉) · M. Stanić
Faculty of Economics and Business in Osijek, Josip Juraj Strossmayer University of Osijek, Osijek, Croatia
e-mail: mirna.leko.simic@efos.hr

M. Stanić
e-mail: marina.stanic@efos.hr

M. Čutura
Faculty of Economics, University of Mostar, Mostar, Bosnia and Herzegovina
e-mail: marija.cutura@ef.sum.ba

I. Pavić et al. (eds.), *Bioethics and Social Ethics in The Modern World*,
https://doi.org/10.1007/978-3-031-86418-6_13

food consumption. The study used a highly structured questionnaire to collect data from 547 respondents. These data were statistically analyzed via descriptive and inferential statistics, including variance and correlation analyses. The results reveal statistically significant differences in the consumption patterns of sustainable food, which are influenced by the level of knowledge about sustainable food. This study contributes to the literature by highlighting the specific behavioral patterns of sustainable food consumption and examining the role that knowledge plays in shaping individual behavior. The findings highlight the need for targeted educational and policy interventions to promote sustainable food consumption practices across different demographic segments.

Keywords Food consumption · Environmental impact · Sustainable behavior · Food preferences

Introduction

Sustainable consumption behavior reflects an understanding of the lasting implications of personal consumption decisions on ecological and social systems (Epstein, 2018). It may be environmentally conscious buying (e.g., choosing green or organic products), practicing energy efficiency, reducing waste and recycling, or engaging in nature-related activities (Tan & Lau, 2011). Čutura and Leko Šimić (2024) identify waste reduction and informed buying as key focuses of contemporary research on sustainable consumption. However, in their bibliometric analysis of SDG12 studies in Central and Eastern Europe, Alfirević et al. (2023) reported that this topic has only recently become of scientific interest in this geographical area and includes areas such as environmentally friendly consumer attitudes and behaviors, environmentally friendly tourism, and, more recently, the circular economy. In addition, the research found that consumers may be willing to adopt eco-friendly technological products even if they offer lower performance, pay higher prices, and change their preferred retail channels to acquire such environmentally responsible goods (Leonidou et al., 2010). In another bibliometric study on sustainable food consumption (Kristia et al., 2023), related topics were particularly relevant and underdeveloped, indicating a significant research gap.

Food production and consumption are often referred to as crucial factors influencing environmental sustainability. The UN 2030 Agenda for Sustainable Development mentions sustainable food consumption as a general goal under SDG 12—Ensuring sustainable consumption and production; SDG 2—Zero Hunger; and SDG 3—Health. As household food consumption contributes to more than 60% of global greenhouse gas emissions (Ivanova et al., 2016), sustainable food purchasing behavior and consumption are becoming increasingly important in practice and scientific research.

Sustainable food consumption is a consumption pattern that minimizes the negative impact on the environment, society, and economy. It is related to the purchase, use, and disposal of food (Razum et al., 2017). It contributes to the United Nations SDGs by minimizing food waste, implementing energy-efficient food preparation processes, and supporting local food production.

Households, i.e., individual consumers, contribute the most to food waste in the EU—over 53% of total food waste (Ilakovac et al., 2020). According to a report by the Croatian Agency for Environment and Nature and the Ministry of Economy and Sustainable Development (2018), Croatian households generate approximately 75 kg of food waste per capita per year, whereas the EU average is 92 kg. At the same time, the share of recycled biowaste (kg per capita) in Croatia is only 12, whereas it is 32 in Hungary, 84 in Spain, and 85 in Portugal, for example (Eurostat, 2020). A recent study (Hendal, 2024) on meat consumption patterns in Croatia reported that 25% of the Croatian population aged 16 and over consumes meat daily, which is an increase of 5% compared with that in 2022, whereas sustainability reports and recommendations suggest a decrease in meat consumption as a way to improve sustainablity in food consumption.

Considering these numbers, where food waste is only one of several elements of unsustainable food consumption, one could conclude that Croatian consumers do not behave sustainably and that there is still much to improve on the consumer side. On the other hand, several studies over the years in Croatia point to a strong preference for local food in the form of ethnocentrism, which is one of the components of sustainable food consumption (Knežević et al., 2018; Leko Šimić, 2000; Tomić, 2016; Živoder, 2022). The same phenomenon has been reported in several other countries (Lončarić et al., 2018). These results illustrate the complexity of evaluating sustainable food consumption.

The findings from the early 2016 regional survey, which included Bosnia and Herzegovina, indicate a relatively low level of food waste among the population (Vaško et al., 2020). Since most food purchases are made in large retail stores, the average person throws away less than half a kilogram of food per week, which costs less than 5 euros. This suggests that food waste is not yet a significant problem due to traditional habits and perhaps also due to the prevailing economic conditions in the region. A recent study (Djekic et al., 2019) revealed that households discard approximately 2.78 kg of food waste per week or 0.95 kg per household member, meaning that approximately 145 kg of food waste is discarded annually, representing a significant increase within a few years.

Moreover, food consumption patterns are notoriously difficult to change, as they are closely linked to lifestyle and highly sociocultural (Vermeir et al., 2020). There is a wide range of social, cognitive, affective, and environmental influences on food consumption patterns at the individual and national levels (Bublitz et al., 2010). One of these elements is also knowledge about sustainable food consumption.

Knowledge is critical in promoting sustainable food consumption by influencing consumer behavior. Knowledge implies understanding and is usually demonstrated through action (Kuzniar et al., 2021). Although not explicitly related to the field of food consumption, several studies suggest that increased knowledge about the environment and sustainability has a significant and positive effect on consumers' pro-environmental behavior (Diamantopoulos et al., 2003; Flamm, 2009; Khan et al., 2020).

This study aims to provide deeper insight into the food consumption patterns of consumers in Croatia and Bosnia and Herzegovina concerning sustainability and to relate these patterns to their knowledge (or lack of knowledge) about sustainable food consumption.

Literature Overview

There are numerous definitions and criteria for sustainable food consumption. Experts agree only on the most general points, namely, that it is the use of food *"that responds to basic needs and brings a better quality of life while minimizing the use of natural resources, toxic materials and emissions of waste and pollutants over the life cycle, so as not to jeopardize the needs of future generation"* (Oslo Roundtable on Sustainable Production & Consumption, 1994). This refers to the increased use of plant- or insect-based foods as well as seasonal, locally produced, and organic foods and the reduction of meat-based foods and no excessive packaging (Vermeir et al., 2020). However, this definition provides only an environmental perspective of sustainability, while the social and economic aspects are missing. With respect to the social aspect, sustainable food refers to human well-being, equality and fairness, justice, social inclusion and participation, and the preservation or conservation of sociocultural characteristics (Jackson & Holm, 2024). In economic terms, sustainable food consumption refers to its accessibility and affordability (Borch & Kjaernes, 2016), i.e., the fair price of products, which should reflect their quality on the one hand and be fair to both the producer and the consumer on the other hand (Pocol et al., 2020). Notably, forecasts show that the average Croatian household will spend 15.4% of the total household budget on food in 2027, which has remained relatively stable since 2009, when the average household expenditure on food amounted to 16.2% of the household budget (Croatia Food & Drink Report—Q4, 2023). In 2024, the average consumer in Bosnia and Herzegovina is forecast to spend approximately USD 1,930 on food and nonalcoholic beverages (Statista, 2024). The latest Household Budget Survey in Bosnia and Herzegovina for the period 2021–2022 shows that households spend 30% of their income on food and soft drinks (Household Budget Survey Bosnia and Herzegovina, 2022). In contrast, according to EUROSTAT, the average household in the European Union spends approximately 13% of household income on identical product categories in the same period (Eurostat, 2022).

In addition to the aforementioned high food waste, low recycling of organic waste, and high meat consumption, Croatian food consumption patterns also indicate deficits in the consumption of fruit, vegetables, and whole grains (Ljubičić et al., 2017), which contradicts both sustainable food consumption as defined by the World Health Organization (*"sustainable diets are rich in vegetables, fruit, and whole grains")* and the need to reduce meat consumption as one of the dimensions of sustainable food consumption patterns.

Looking at these criteria and the current state of global food consumption and its consequences, it is clear that changing food consumption patterns could contribute significantly to more sustainable global growth and development. If consumers are required to change their behavior, they must be well-informed and aware of the impacts of their food consumption patterns, choices, and desirable behaviors in this context (Vermeir et al., 2020). Cerjak et al. (2010) surveyed the urban population buying organic products in the main cities of Croatia, Slovenia, and Bosnia and Herzegovina. Respondents from all three countries expressed favorable attitudes toward organic food. Croatia and Slovenia's primary motives for purchasing organic products include health-related benefits and environmental considerations. Meanwhile, the main drivers center on reconnecting with nature, health value, and food safety in Bosnia and Herzegovina. Overall, this regional study's findings indicate no significant differences in organic food consumption behavior across the three markets.

However, studies on the quality of nutrition paint a somewhat different picture. A study conducted in Bosnia and Herzegovina (Gicevic et al., 2019) revealed that the mean value of the prime diet quality score (PDQS) was 15.8 (ranging from 7–28 out of 42 possible points). These findings suggest that the adult populations in Bosnia and Herzegovina generally have low PDQS scores due to the high consumption of refined grains, high-fat dairy products, and processed meat and low consumption of whole grains, nuts, and fish.

Many studies have recognized education as a key tool for long-term changes in certain behaviors (Kelder et al., 1994; Luepker et al., 1983) in many areas (e.g., alcohol, drugs, smoking, and antiviolence programs). Educating consumers about sustainable food and the environmental, social, and health impacts of their food choices can lead to more informed decisions in both purchasing and consumption.

Knowledge about sustainability and food significantly predicts sustainable food consumption behavior (Kuźniar et al., 2021; Verain et al., 2015). A study by Gram-Hanssen (2010) revealed that consumers with greater environmental knowledge are significantly more willing to participate in environmentally friendly activities. Knowledge was identified as the main difference between consumers who are interested in environmental issues and those who are not (Fransson & Garling, 1999). Interest in environmental issues is related to pro-environmental behavior, which includes sustainable food consumption. Several studies have shown that a lack of information (Gifford, 2011), information overload (Neumann et al., 2012), and confusion (Chen & Chang, 2013) can contribute to the low adoption of sustainable behaviors. A recent study conducted in Croatia (Mesić et al., 2021) revealed that just over one-third of respondents were familiar with sustainable food. The respondents recognized them as products whose production has a lower harmful impact on the environment and are produced according to the principles of organic farming and as locally produced food.

Furthermore, over 80% of respondents who do not buy sustainable food say that they do not know if the food they buy is sustainable. These findings align with those of a previous study (Sokolić et al., 2015), which reported that 61% of Croatian consumers never or rarely read food labels for information, whereas only 10% do so regularly. Furthermore, only 26% of the respondents said that they fully understood the information on food labels.

The literature review clearly links sustainability, environmental and food knowledge, and sustainable food consumption patterns. This shows that considerable efforts should be made to increase knowledge on these topics to promote and motivate positive changes in food consumption toward sustainability.

Methodology

In this study, we analyze the food consumption behavior of Croatian and Bosnian-Herzegovinian consumers with respect to sustainability and relate it to knowledge about sustainable food. We identify the actual food consumption patterns, the general level of knowledge about sustainable food, and the most common "black holes" in this knowledge. The initial assumption is that more knowledge about sustainable food positively correlates with more sustainable food consumption.

Sample Description

The respondents were selected via a nonprobabilistic sampling method. The questionnaire was distributed in a combination of online and paper questionnaires to ensure the inclusion of different demographic groups. The online questionnaire was the predominant method used to reach a large and diverse audience, whereas the paper questionnaire was used to reach elderly individuals with limited internet access and knowledge. Data collection took place over three months (February to April 2024). The sample characteristics are presented in Table 13.1.

Table 13.1 Sample characteristics (N = 547)

Variable	*Category*	*Frequency*	*Percentage*
Gender	Male	183	33.5
	Female	364	66.5
Age (yrs)	18–25	138	25.2
	26–35	87	15.9
	36–45	98	17.9
	46–55	102	18.6
	56–65	84	15.4
	Over 65	38	6.9
Status	Student	131	23.9
	Employed	316	57.8
	Business owner	31	5.7
	Retired	56	10.2
	Unemployed	13	2.4
Education level	Elementary school	7	1.3
	High school	185	33.8
	University	355	64.9
Living standard	Below average	25	4.6
	Average	435	79.5
	Above average	87	15.9
Country of origin	Croatia	346	63.3
	Bosnia and Herzegovina	201	36.7

Source: Authors

Variables and Methods

To assess respondents' knowledge of sustainable food, the authors developed a set of questions on the basis of a thorough analysis of previous studies. The questions were presented in a true-or-false format, with an additional "I do not know" option for each statement. The questionnaire consists of seven statements, which are shown in Table 13.2. Statements two and six were false, whereas all others were true.

To enable different statistical analyses, the variable "knowledge about sustainable food" was operationalized in two ways. The first version of the knowledge variable was calculated as the sum of the correct answers (where the minimum value of the variable was zero and the maximum value was seven). The descriptive analysis yielded a mean value of 5.13, with a standard deviation of 1.24. This variable was used in the correlation analysis. The second version of the knowledge variable was defined as an ordinal categorical variable with three categories: 1—respondents with good knowledge of sustainable food (respondents who gave five,

Table 13.2 Dimensions of sustainable food knowledge construct

Statement No	*Statement*
S1	Food that is grown or produced locally or regionally
S2	Food packaged in such a way that it is visible (e.g., in transparent packaging)
S3	Food that is high quality, safe, and healthy
S4	Food that enables the maintenance of biodiversity
S5	Food that contributes to the welfare of producers
S6	Vegan food
S7	Food that reduces the impact on the environment

Source: Authors

six, or seven correct answers), 2—respondents with sufficient knowledge of sustainable food (respondents who gave three or four correct answers), and 3—respondents with insufficient knowledge of sustainable food (respondents who gave only one or two correct answers or for whom all answers were incorrect).

The focus variable used in the study measures a person's sustainable food consumption behavior. As presented in Table 13.3, the measurement scale consisted of six items that depicted different behavioral patterns concerning environmental impact (adapted from Tobler et al., 2011) and food waste (adapted from Leonidou et al., 2010) as well as health impact and eating habits (adapted from Tobler et al., 2011). The degree of agreement with the statements on the scale was measured on a Likert scale from 1 to 5.

Results and Discussion

We began the analysis by assessing the respondents' behavioral patterns concerning sustainable food consumption. Although theory identifies the environmental, economic, and social impacts of sustainable food consumption patterns (Borch & Kjærnes, 2016; Jackson & Holm, 2024; Reisch, Eberle & Lorek, 2013), this study goes a step further by identifying specific behavioral patterns, such as local and seasonal food preferences, food waste reduction, and health impacts, which are increasingly becoming the focus of current research on sustainable food consumption (Ran et al., 2022).

Table 13.3 Dimensions of the sustainable food consumption construct

Dimension	*Items*	*Source*
Environmental impact	Avoid consumption of any imported agricultural products and foodstuffs	Tobler et al. (2011)
Seasonal food preferences	Eat primarily seasonal fruits and vegetables	Tobler et al. (2011)
Food waste reduction	Reduce household food waste whenever possible Recycle household food waste whenever possible	Leonidou et al. (2010)
Health impact	Eat less meat, maximum once or twice per week Eat plant-based meat substitutes	Tobler et al. (2011)

Source: Authors

On the basis of previous studies and the aim of this research, four distinct consumer behaviors were measured (see Table 13.4).

As highlighted in Fig. 13.1, the most pronounced sustainable food consumption behavior is related to choosing seasonal foods and reducing food waste whenever possible. On the other hand, the impact of sustainable food consumption on health was the least pronounced. This contrasts with the recent literature review by Ran et al. (2022), which revealed that health awareness and concerns were the most frequently cited factors influencing sustainable food consumption behaviors in numerous studies, whereas environmental impacts were significantly less important. On the other hand, this could be understandable, as we included elements in the health impact measurement opposite to the traditional dietary habits in both countries, characterized as cultures with high meat consumption.

Table 13.4 Behavioral patterns of sustainable food consumption

Dimension	*Number of items*	*Mean*	*St Dev*
Environmental impact	1	3.21	1.157
Seasonal food preferences	1	4.14	0.913
Food waste reduction	2	4.04	0.933
Health impact	2	2.23	1.028

Source: Authors

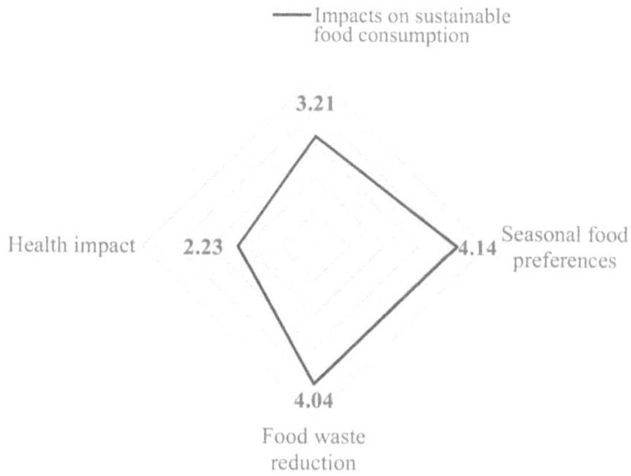

Fig. 13.1 Sustainable food consumption patterns (Source: Authors)

This is a strong tradition, and traditions are associated with less sustainable consumption behavior (Maciejewski, 2020). Furthermore, habitual behavior prevails for products such as low-involvement food (Jager, 2000).

The next step was to assess the respondents' familiarity with various aspects of sustainable food. Their knowledge was assessed on seven true and false statements. The statements were extracted and adapted from the definitions of sustainable food in the abovementioned studies. Finally, respondents were divided into three categories on the basis of their knowledge of sustainable food: good knowledge (five to seven correct answers), sufficient knowledge (three or four correct answers), and insufficient knowledge (zero to two correct answers). The vast majority of respondents (73.5%) were aware of the characteristics of sustainable food, as shown in Table 13.5. This finding is somewhat unexpected, as a study from just three years ago (Mesić et al., 2021) revealed a significantly lower level of knowledge. The high level of knowledge in this study could be related to the characteristics of the sample, in which 64.9% of the respondents had a higher education degree. A Slovakian study (Holotová et al., 2021) confirmed a significant correlation between education level and information seeking concerning sustainable food products, suggesting that consumers with higher levels of education base their food consumption decisions on information.

Education and information measures that lead to knowledge creation are commonly applied behavior change interventions used to change

Table 13.5 Respondents' knowledge about sustainable food (total)

*Category**	*Frequency*	*Percentage*
Good	402	73.5
Sufficient	127	23.2
Insufficient	18	3.3

* Good (5, 6 or 7 correct answers), Sufficient (3 or 4 correct answers), Insufficient (0, 1 or 2 correct answers)
Source: Authors

consumer behavior. However, the actual effectiveness of information and education in influencing long-term consumer behavior change is usually not recorded and thoroughly evaluated and is therefore not clearly understood. (Temme et al., 2020). A study by Grilli and Curtis (2021) indicated that despite their frequent use in marketing research, education, and awareness-raising methods have the lowest success rate in changing consumer behavior toward more sustainable food consumption. Andreasen (1995) also warned that education alone cannot be expected to bring about sustainable behavior, as it influences only awareness. Education does not necessarily change attitudes, feelings, or control behavior. Instead, marketing efforts that identify and leverage behavioral levers have the best chance of success. Several studies have shown that information and education are most effective when combined with other strategies, such as marketing, as White et al. (2019) noted.

Table 13.6 shows the specific knowledge about sustainable food. Almost all the respondents recognized sustainable foods as those that are of high quality, safe, and healthy (96.1%), that contribute to biodiversity conservation (86.6%), and that are grown locally or regionally (81%). On the other hand, 43.6% of respondents believe that transparent packaging is part of the definition of sustainable food, and 24.2% do not know that sustainable food contributes to the well-being of producers. Interestingly, 18.5% of the respondents did not know whether vegan food is sustainable.

Further analysis revealed that food consumption preferences and behaviors vary significantly depending on the level of knowledge about sustainable food. One-way ANOVA revealed a statistically significant difference in food consumption preferences across knowledge categories. In particular, individuals with good knowledge about sustainable food are significantly more likely to consume seasonal food than are individuals with sufficient knowledge about sustainable food ($F(2, 544) = 5682$, p

Table 13.6 Respondents' knowledge about sustainable food (by questions)

Statement No	*Correct*	*Correct (%)*	*Incorrect*	*Incorrect (%)*	*Do not know*	*Do not know (%)*
S1 (locally produced)	442	81.0	60	11.0	44	8.1
S2 (transparent packaging)	241	44.1	**238**	**43.6**	67	12.3
S3 (quality and safety)	**524**	**96.1**	14	2.6	7	1.3
S4 (saving biodiversity)	473	86.8	18	3.3	54	9.9
S5 (producers' well-being)	330	60.4	132	24.2	84	15.4
S6 (vegan)	373	68.2	72	13.2	**101**	**18.5**
S7 (environmental impact)	424	77.9	63	11.6	57	10.5

Source: Authors

Table 13.7 Correlation analysis

Variable	*(1)*	*(2)*	*(3)*	*(4)*	*(5)*
(1) Knowledge about sustainable food	1				
(2) Environmental impact	,150**	1			
(3) Seasonal food preferences	,128**	,350**	1		
(4) Food waste reduction	,140**	,337**	,472**	1	
(5) Health impact	,040	,365**	,081	,214**	1

* Correlation is significant at the 0.05 level
** Correlation is significant at the 0.01 level
Source: Authors

< 0.05). In addition, individuals with good knowledge about sustainable food tend to avoid imported food to a significantly greater extent than do individuals with sufficient and insufficient knowledge ($F(2, 544) = 6740$, $p < 0.001$).

The correlation analysis examined the strength and direction of the relationship between the level of knowledge about sustainable food and four dimensions (or behavioral patterns) of sustainable food consumption. Knowledge about sustainable food has a statistically significant but weak relationship with three behavioral patterns. However, the results show that some positive behaviors are interrelated, and it can be assumed that one type of positive behavior favors the other. This applies particularly to the relationship between seasonal food consumption and waste reduction.

Previous research has shown that food consumption behavior is more complex and challenging to change than food waste behavior is (Redman, 2013), which is also confirmed in the study by Attiq et al. (2021), which shows that consumers with higher levels of environmental knowledge are likely to engage in food waste reduction behavior but not consumption behavior.

Conclusion

This study aimed to gain deeper insight into food consumption patterns from a sustainability perspective and to relate these patterns to knowledge (or lack of knowledge) about sustainable food consumption. Choosing seasonal foods and reducing food waste were the most common characteristics of sustainable food consumption patterns, whereas the impact of food consumption on health was the least pronounced. Almost three-fourths of the respondents had good knowledge of sustainable food.

Knowledge can influence consumer behavior in various ways. First, it raises awareness of food production and consumption's environmental, social, and health impacts. If consumers understand the consequences of unsustainable food choices, they may be more motivated to change their habits. Second, it helps them make more informed choices about the food they buy and consume, e.g., to recognize sustainable products and understand the impact of food on health, society, and the environment. Providing accurate, accessible, and relevant information on sustainable food practices and their benefits can empower consumers to choose more sustainable food, thus contributing to a more sustainable food system.

This study's main contribution is identifying the sustainable food consumption patterns of Croatian and Bosnian-Herzegovinian consumers and the role of knowledge about sustainable food in their consumption behavior. As mentioned above, knowledge alone cannot fully explain behavior; several other factors need to be considered, such as socioeconomic and cultural factors and personal attitudes and values. The limitations of this study relate primarily to the specifics of the sample. The high proportion of respondents with a university degree presumably led to an above-average general knowledge of sustainable food consumption, which is also reflected in the behavioral patterns. In addition, the fact that the sample was predominantly made up of young respondents could have influenced variables such as the impact on health (in the sense that young people do not pay as much attention to their health as the older population does). This ultimately limits the possibility for generalization. Finally, the results of this study may not be comparable to those of other studies in the field, which have used different measurement scales to assess the most important variables. This fragmentation hinders development, and future studies should address this issue by examining and evaluating the primary measurement constructs commonly used in this field of research.

References

Alfirević, N., Zdrilić, I. & Rendulić, D. (2023). Mapping the sustainable production and consumption literature in South East Europe. In: Tufek-Memišević, T., Arslanagić-Kalajdžić, M., & Ademović, N. (Eds.) *Interdisciplinary Advances in Sustainable Development*. ICSD 2022. Lecture Notes in Networks and Systems, vol. 529. Springer. https://doi.org/10.1007/978-3-031-17767-5_10

Andreasen, A. (1995). Marketing social change: changing behavior to promote health, social development and the environment. Jossey—Bass Publishers.

Attiq, S., Habib, M. D., Kaur, P., Hasni, M. J. S., & Dhir, A. (2021). Drivers of food waste reduction behavior in the household context. *Food Quality and Preference, 94*, 104300. https://doi.org/10.1016/j.foodqual.2021.104300

Borch, A., & Kjærnes, U. (2016). Food security and food insecurity in Europe: An analysis of the academic discourse (1975–2013). *Appetite, 103*, 137–147. https://doi.org/10.1016/j.appet.2016.04.005

Bublitz, M. G., Perrachio, R. A., & Block, R. G. (2010). Why did I eat that? Perspectives on food decision-making and dietary restraint. *Journal of Consumer Psychology, 20*(3), 239–258. https://doi.org/10.1016/j.jcps.2010.06.008

Cerjak, M., Mesić, Ž, Kopić, M., Kovačić, D., & Markovina, J. (2010). What motivates consumers to buy organic food: Comparison of Croatia, Bosnia Herzegovina, and Slovenia. *Journal of Food Products Marketing, 16*(3), 278–292. https://doi.org/10.1080/10454446.2010.484745

Chen, Y.-S., & Chang, C.-H. (2013). Greenwash and green trust: The mediation effects of green consumer confusion and green perceived risk. *Journal of Business Ethics, 114*(3), 489–500. https://doi.org/10.1007/s10551-012-1360-0

Čutura, M., & Leko Šimić, M. (2024). *Društvena strana marketinga*. Pressum.

Diamantopoulos, A., Schlegelmilch, B. B., Sinkovics, R. R., & Bohlen, G. M. (2003). Can sociodemographics still play a role in profiling green consumers? A review of the evidence and an empirical investigation. *Journal of Business Research, 56*(6), 465–480. https://doi.org/10.1016/S0148-2963(01)00241-7

Djekic, I., Operta, S., Djulancic, N., Lorenzo, J. M., Barba, F. J., Djordjević, V., & Tomasevic, I. (2019). Quantities, environmental footprints and beliefs associated with household food waste in Bosnia and Herzegovina. *Waste Management & Research, 37*(12), 1250–1260. https://doi.org/10.1177/0734242X19873709

Epstein, M. J. (2018). Making sustainability work: Best practices in managing and measuring corporate social, environmental and economic impacts. *Routledge*. https://doi.org/10.4324/9781351280129

Eurostat. (2020). *Energy productivity (SDG_07_30) indicator profile*. Eurostat: Luxembourg.

Eurostat. (2022). Euro Area - Final consumption expenditure of households: Food and nonalcoholic beverages. Retrieved from: https://tradingeconomics.com/euro-area/final-consumption-expenditure-of-households-food-non-alcoholic-beverages-eurostat-data.html

Fitch Solutions Country Industry Reports (2023). *Croatia Food & Drink Report - Q4 2023*. Retrieved from https://www.proquest.com/docview/2854659946/fulltextPDF/C14985A63BEA4461PQ/21?accountid=44407&sourcetype=Reports

Flamm, B. (2009). The impacts of environmental knowledge and attitudes on vehicle ownership and use. *Transportation research part D: Transport and environment* Fransson, N., Garling, T. (1999). Environmental concern: Conceptual definitions, measurement methods and research findings. *Journal of Environmental Psychology, 19*, 369–382. https://doi.org/10.1006/jevp.1999.0141

Gicevic, S., et al. (2019). Demographic and socioeconomic predictors of diet quality among adults in Bosnia and Herzegovina. *Public Health Nutrition, 22*(17), 3107–3117. https://doi.org/10.1017/S1368980019001988

Gifford, R. (2011). The dragons of inaction: Psychological barriers that limit climate change mitigation and adaptation. *American Psychologist, 66*(4), 290–302. https://doi.org/10.1037/a0023566

Gram-Hanssen, K. (2010). Standby consumption in household analyzed with a practice theory approach. *Journal of Industrial Ecology, 14*, 150–165. https://doi.org/10.1111/j.1530-9290.2009.00194.x

Grilli, G., & Curtis, J. (2021). Encouraging pro-environmental behaviors: A review of methods and approaches. *Renewable and Sustainable Energy Reviews, 135*, 110039. https://doi.org/10.1016/j.rser.2020.110039

Hendal, Ja trgovac (2024). Istraživanje prehrambenih navika hrvatskih potrošača vezanih uz potrošnju mesa. Retrieved from https://jatrgovac.com/jatrgo 2018/meso-i-mesne-preradevine-cijena-presudno-utjece-na-konzumaciju/

Holotová, M., Horská, E., & Nagyová, Ĺ. (2021). Changing patterns of sustainable food consumption regarding environmental and social impact-insights from Slovakia. *Frontiers in Sustainable Food Systems, 5*, 703827. https://doi.org/10.3389/fsufs.2021.703827

Ilakovac, B., Voca, N., Pezo, L., & Cerjak, M. (2020). Quantification and determination of household food waste and its relation to sociodemographic characteristics in Croatia. *Waste Management, 102*, 231–240. https://doi.org/10.1016/j.wasman.2019.10.042

Ivanova, D., Stadler, K., Steen Olsen, K., Wood, R., Vita, G., Tukker, A., et al. (2016). Environmental impact assessment of household consumption. *Journal of Industrial Ecology, 20*, 526–536. https://doi.org/10.1111/jiec.12371

Jackson, P., & Holm, L. (2024). Social and economic dimensions of food sustainability—A background paper for the Nordic Nutrition Recommendations. *Food & Nutrition Research, 5*, 68. https://doi.org/10.29219/fnr.v68.10450

Jager, W. (2000). *Modeling consumer behavior*, PhD thesis, University of Gronningen. Retrieved from https://pure.rug.nl/ws/portalfiles/portal/991 4457/c5.pdf

Kelder, S. H., Perry, C. L., Klepp, K.-I., & Lytle, L. L. (1994). Longitudinal tracking of adolescent smoking, physical activity, and food choice behaviors. *American Journal of Public Health, 84*(7), 1121–1126. https://doi.org/10.2105/AJPH.84.7.1121

Khan, M. S., Saengon, P., Alganad, A. M. N., Chongcharoen, D., & Farrukh, M. (2020). Consumer green behavior: An approach toward environmental sustainability. *Sustainable Development, 28*(5), 1168–1180. https://doi.org/10.1002/sd.2066

Knežević, B., Orban, M., Škrobot, P., & Delić, M. (2018). Perception of traditional food products and consumer ethnocentrism: Evidence from Croatia. *Trade Perspectives, 2018*, 87.

Kristia, K., Kovacs, K., Bacs, Z., & Rabbi, M. F. (2023). A bibliometric analysis of Sustainable food consumption: Historical evolution, dominant topics and trends. *Sustainability, 23*(15), 8998. https://doi.org/10.3390/su15118998

Kuźniar, W., Surmacz, T., & Wierzbiński, B. (2021). The impact of ecological knowledge on young consumers' attitudes and behaviors toward the food market. *Sustainability, 13*(4), 1984. https://doi.org/10.3390/su13041984

Leko-Šimić, M. (2000). Hrvatsko gospodarstvo u uvjetima otvorenog tržišta uspješnost inozemnih proizvoda na hrvatskom tržištu i marketing implikacije za hrvatsko gospodarstvo. In *Tržišna demokracija u Hrvatskoj stanje i perspektive* , 209–219.

Leonidou, L. C., Leonidou, C. N., & Kvasova, M. O. (2010). Antecedents and outcomes of environmentally friendly consumer attitudes and behavior. *Journal of Marketing Management, 26*(13–14), 1319–1344. https://doi.org/10.1080/0267257X.2010

Lončarić, R., Sudarić, T., & Deže, J. (2018). Consumer ethnocentrism in the global food market. *Interdisciplinary Management Research XIV*, *618*.

Luepker, R., Johnson, C. A., Murray, D., & Pechacek, T. (1983). Prevention of cigarette smoking: Three-year follow-up of an education program for youth. *Journal of Behavioral Medicine, 6*(1), 53–62. https://doi.org/10.1007/BF00845276

Ljubičić, M., Matek Sarić, M., Colić Barić, I., Rumbak, I., Komes, D., Šatalić, Z., & Guiné, R. P. (2017). Consumer knowledge and attitudes toward healthy eating in Croatia: A cross-sectional study. *Arhiv Za Higijenu Rada i Toksikologiju, 68*(2), 153–158. https://doi.org/10.1515/aiht-2017-68-2939

Maciejewski, G. (2020). Consumers toward sustainable food consumption. *Marketing of Scientific and Research Organizations, 36*(2), 19–30. https://doi.org/10.2478/minib-2020-0014

Mesić, Željka, Glaser, Tara, & Tomić Maksan, Marina (2021). Održivi prehrambeni proizvodi—potrošačke navike i stavovi hrvatskih potrošača (Sustainable food products—habits and attitudes of Croatian consumers). In Book of Abstract of 1st international conference Food and Climate change/-Koprivnica, Hrvatska, 15 - 16.10.2021, 3434.

Ministry of economy and sustainable development (2018). *Improving the system for collecting data on biowaste and food waste*, Zagreb, Croatia.

Neumann, B. R., Roberts, M. L., & Cauvin, E. (2012). management control systems dilemma: Reconciling sustainability with information overload. In Marc J. Epstein & John Y. Lee (Eds.). *Advances in Management Accounting*. Emerald Group, 1–28. https://doi.org/10.1108/S1474-7871(2012)0000020007

Oslo Roundtable on Sustainable Production and Consumption. (1994). The imperative on sustainable production and consumption. Retrieved from https://enb.iisd.org/topics/sustainable-consumption-production

Pocol, C. B., Marinescu, V., Amuza, A., Cadar, R. -L., & Rodideal, A. A. (2020). Sustainable vs. unsustainable food consumption behavior: A study among students from Romania, Bulgaria and Moldova. *Sustainability, 12*(11), 4699. https://doi.org/10.3390/su12114699

Ran, Y., Lewis, A. N., Dawkins, E., Grah, R., Vanhuyse, F., Engström, E., & Lambe, F. (2022). Information as an enabler of sustainable food choices: A behavioral approach to understanding consumer decision-making. *Sustainable Production and Consumption, 31*, 642–656. https://doi.org/10.1016/j.spc.2022.03.026

Razum, A., Pandža Bajs, I., & Zekić, Z. (2017). Analiza čimbenika održive potrošnje generacije Z u modnoj industriji. *Ekonomski Pregled, 68*(3), 297–318.

Reisch, L., Eberle, U., & Lorek, S. (2013). Sustainable food consumption: an overview of contemporary issues and policies. *Sustainability: Science, Practice and Policy*, 9(2), 7–25. https://doi.org/10.1080/15487733.2013.11908111

Sokolić, D., Kolarić Kravar, S., Batinić Sermek, M., & Colić Barić, I. (2015). *Razumijevanje informacija na hrani i stav potrošača o bacanju hrane*, HAH.

Statista. (2024). Consumption Indicators - Bosnia and Herzegovina. Retrieved from https://www.statista.com/outlook/co/consumption-indicators/bosnia-and-herzegovina

Tan, B.-C., & Lau, T. C. (2011). Green purchase behavior: Examining the influence of green environmental attitude, perceived consumer effectiveness and specific green purchase attitude. *Australian Journal of Basic and Applied Sciences, 5*(8), 559–567.

Temme, E. H. M., Vellinga, R. E., de Ruiter, H., Kugelberg, S., van de Kamp, M., Milford, A., Alessandrini, R., Bartolini, F., Sanz-Cobena, A., & Leip, A. (2020). Demand-side food policies for public and planetary health. *Sustainability, 12*(15), 5924. https://doi.org/10.3390/su12155924

The Household Budget Survey Bosnia and Herzegovina. (2022). Retrieved from https://bhas.gov.ba/data/Publikacije/Saopstenja/2023/CIS_01_2022_Y1_1_BS.pdf

Tobler, C., Visschers, V. H., & Siegrist, M. (2011). Eating green. Consumers' willingness to adopt ecological food consumption behaviors. *Appetite, 57*(3), 674–682. https://doi.org/10.1016/j.appet.2011.08.010

Tomić, M. (2016). *Application of theory of planned behavior in ethnocentric buying behavior*. Doctoral thesis. University of Zagreb, Faculty of Agriculture, Zagreb

Vaško, Ž., Berjan, S., El Bilali, H., Allahyari, M. S., Ostojić, A., Bottalico, F., ... & Capone, R. (2020). Attitude and behavior of Bosnian households toward food waste. *Agriculture & Forestry/Poljoprivreda i šumarstvo*, *66*(4), 139–150. https://doi.org/10.17707/AgricultForest.66.4.11

Verain, M. C., Dagevos, H., & Antonides, G. (2015). Sustainable food consumption. Product choice or curtailment? *Appetite, 91*, 375–384. https://doi.org/10.1016/j.appet.2015.04.055

Vermeir, I., et al. (2020). Environmentally sustainable food consumption: A review and research agenda from a goal-directed perspective. *Frontiers in Psychology, 11*, 520238. https://doi.org/10.3389/fpsyg.2020.01603

White, K., Habib, R., & Hardisty, D. J. (2019). How to SHIFT consumer behaviors to be more sustainable: A literature review and guiding framework. *Journal of Marketing, 83*(3), 22–49. https://doi.org/10.1177/0022242919825649

Živoder, D. (2022). *Marketinške implikacije označavanja domaćih poljoprivrednih i prehrambenih proizvoda putem prepoznatljivih oznaka i zaštićenih znakova (Doctoral dissertation*. Josip Juraj Strossmayer University of Osijek. Faculty of Agricultural Sciences Osijek.

CHAPTER 14

Dalmatian Cuisine: A Traditional Facilitator of Ethics and Sustainability

Vedran Poljak, Mateja Dugandžić, Eva Pavić, and Valentina Rahelić

Abstract Traditional cuisine is Dalmatia's diverse, nutritionally favorable, and unique cultural heritage. It is primarily defined by simple ingredients from the Mediterranean biome—olive oil, wine, bread, seafood, and local fauna. The food preparation is simple and healthy, with a wealth of interesting and diverse flavors of herbs. Its development reflects the region's turbulent history, shaped by encounters between ancient Greeks and Romans, as well as conflicts between the Levant and the West. Dalmatian cuisine preserves ancestors' stories in unique dishes and the customs

V. Poljak (✉)
University Department of Marine Studies, University of Split, Split, Croatia
e-mail: vpoljak@unist.hr

M. Dugandžić
Study of Hotel Management and Gastronomy in Makarska, University of Split, Makarska, Croatia

E. Pavić · V. Rahelić
Department of Nutrition and Dietetics, University Hospital Center, Zagreb, Croatia

I. Pavić et al. (eds.), *Bioethics and Social Ethics in The Modern World*,
https://doi.org/10.1007/978-3-031-86418-6_14

related to those dishes. Its sustainability stems from Mediterranean agricultural practices, which involve resourceful methods of cultivation on dry, rocky terrain with limited arable land. Since the dawn of humanity, *hortus* (garden) and *ager* (field) have provided sustenance for local inhabitants, and *silva* (forest) supplied the wood for their hearths. The sea, salt pans, and fishing have a thousand-year written history and occupy an important position in the Dalmatian menu, expanding the range of dishes to unimaginable limits. Importantly, salty and sweet dishes conquer the palate equally well—the flavors of sour and bitter permeate in a unique harmony. Although traditional Dalmatian cuisine is unique, it has remained open to discoveries from the New World, and today, potatoes and tomatoes have become indispensable. Over many centuries, the Dalmatians, *cum grano salis*, learned to live with their environment as their only home and took from it only as much as needed, preserving nature for future generations.

Keywords Tradition · Cuisine · Dalmatia · Food · Sea · Wine · Olive oil

Introduction

Every country has a characteristic style of gastronomy, such as its national language and traditional music, which are part of its national culture. Croatia has emerged out of the three kingdoms. It does not have one characteristic cuisine for the country but is divided into several distinct cuisines. Each region has its own distinctive dishes and culinary tradition, which is remarkable for such a small country (Vitez & Muraj, 2001).

People often honor their country for sculptors, writers, and poets; we should include gastronomes in this recognition, as preparing a true local dish involves thousands of years of accumulated knowledge and innovation (Calussi, 1997). As the name suggests, Dalmatian cuisine is related to the broader area of the Dalmatian region. Dalmatia is spread over 12,000 kilometers of the Eastern Adriatic coast and has approximately one thousand islands. Although the sea and sun entice visitors to Dalmatia for relaxation, local cuisine provides equally memorable experiences. Since Dalmatian cuisine is related to an extensive regional area, it can be divided into different food preparation methods in coastal, island, and inland areas. Compared with coastal and island regions, Dalmatian

Zagora is relatively modest, relying primarily on cabbage, potatoes, and corn, especially for making polenta. Historically limited resources in this region have shaped a distinctive cooking style.

On the coast, diet is aligned with what the sea and soil provide. There is much Italian, Turkish, and French influence on the coast. The trade ties with the Levant left a deep mark on gastronomy. The island of Dalmatia is focused mainly on the sea, i.e., fish and seafood. The fish were always prepared in various ways: salted, marinated, and even dried. The islands' interior is oriented toward animal husbandry, so lamb dishes are also famous in addition to fish. All three regions are characterized by the addition of different spices to dishes. Zagora is symbolized by a sage, the coastal part by a rosemary, and the islands by an anemone. Dalmatian cuisine combines Mediterranean influences permeated with sea and island cuisine and the culinary peculiarities of the Dalmatian Zagora. Each dish has its own unique story and attracts people, in a special way, to taste it again (Marjanović Radica, 1976). Dalmatian cuisine is an example of the Mediterranean diet, which is related to the way, culture, and way of life of the Mediterranean climate. The Mediterranean diet and way of life of southern European countries and their link with health were first described by Ancel Keys (Montani, 2021).

It is characterized by a high intake of fruits, leafy vegetables, legumes, fish and fish products, nuts, seeds, whole grains, dairy products, and olive oil and a low intake of bakery and baked goods, soft drinks, and red and processed meat (Agnoli et al., 2018; Tosti et al., 2018). These vegetables, fruits, olive oil, and wine are significant sources of various polyphenols with intense antioxidant activity, which is attributed to the significant positive effects of the Mediterranean diet on health (Trichopoulou & Vasilopoulou, 2000). Additionally, the Mediterranean diet is a source of essential nutrients such as fiber, vitamins, minerals, phytochemicals, and other bioactive compounds that can impact long-term sustainable health and prevent the development of chronic noncommunicable diseases (Zong et al., 2016). In Dalmatia, meals are strictly determined by the seasonality of the year and national or religious holidays accompanied by particular food. They are mostly folk social events with abundant meals rich in different vegetables and spices, seeds, and cheeses, with some meat or fish as preferred protein sources. Dalmatia has many written and unwritten customs that people mostly follow. Customs are concerned primarily with holidays, and holidays are considered to maintain gastronomic and cultural customs. The smells of cods and fritters, Christmas tree decorations, and fasting are indispensable Christmas Eve customs in Dalmatia (Calussi, 1997).

Seasonal fruit is a typical dessert in Dalmatian cuisine, whereas nuts and olives are essential snacks. Adopted agricultural knowledge from the ancient Greeks, the division into *hortus* (garden) and *ager* (field), which feed Dalmatian man, is still present today. At the same time, *silva* (forest) keeps fire burning in Dalmatian hearthstone. The sea, salt pans, and fishing have a thousand-year written history and occupy an important position in the Dalmatian menu, expanding the range of dishes to unimaginable limits. Importantly, both salty and sweet dishes conquer the palate equally well—the flavors of sour and bitter permeate in a unique harmony of dishes. Although traditional Dalmatian cuisine is unique, it has remained open to discoveries in the New World, and today, potatoes and tomatoes have become indispensable (Calussi, 2008).

Vineyards and wines are indispensable parts of Dalmatia. This is also one of the reasons why people fall in love with it, in addition to its unique cuisine and landscape. Without wine, food would not be as good as it is with it. In addition to the fact that Dalmatia is known for its food, it is also known as the Croatian wine region. Wines and olive oil are suitable for the Mediterranean climate because of their hot, dry summers and mild winters. Central and southern Dalmatia are mostly stranded out for exploring and tasting wine. Authentic Dalmatian wines create a memorable experience for every tourist, and dinner without a glass of it is unthinkable. The development of wine tourism contributes to the positioning and recognition of a particular tourist area and creates a competitive advantage. The main reasons for the arrival of tourists in Dalmatia (TOMAS, 2020) are passive relaxation (75%), entertainment (44%), new experiences and adventures (24%), gastronomy (22%), and cultural sights and events (7%). Since there are very few data about guests' perceptions of our cuisine, this research aimed to determine guests' knowledge of Dalmatian cuisine and evaluate it.

Materials and Methods

We collected data in person via a self-designed survey questionnaire administered to a simple random sample of participants, with a focus on their attitudes and opinions regarding local food products and dishes. A sample size of 100 respondents was used to determine the perception of Dalmatian cuisine among guests who visited Split-Dalmatia County. The survey was conducted in August 2023 by trained interviewers. Since the survey questionnaire was simple and understandable, the survey lasted only a few minutes. The results are presented via descriptive statistics.

Results

The survey results revealed that the subjects were between 20 and 70 years old, with the dominance of the 30–40 years age group (34%). The gender distribution was as expected, with a predominance of women (57%) compared with men (43%). Approximately one-third (30%) of the respondents stated that they were in Dalmatia for the first time, and two-thirds stated that they were in Dalmatia for the second time or more (70%). The majority of surveyed guests (54%) stated that Dalmatian cuisine is an important reason for visiting Dalmatia. In comparison, another 13% stated that gastronomy is a possible reason for this. One-third of the respondents (33%) stated that gastronomy is not the reason for coming to this destination, which is shown in Fig. 14.1.

Figure 14.2 shows the traditional Dalmatian dishes that guests reported tasting.

Figure 14.2 shows that more than one-fourth of the visitors (28%) tasted "fritule" Croatian mini doughnuts and, out of the main dishes, "pašticada" (Dalmatian beef stew—22%). Meat dishes predominated over "brudet" dishes (Croatian fish stew—9%) and black risotto dishes (5%). Only 2% of the respondents tasted the traditional dish "soparnik" (Dalmatian chard pie).

In the complete assessment of Dalmatian cuisine, most respondents rated it as good (47%) or very good (20%), as shown in Fig. 14.3.

Figure 14.3 clearly shows that two-thirds of guests rated Dalmatian cuisine as good or very good (67%). When the excellent rating is added (17%), 84% of guests rated Dalmatian cuisine with a rating from good to

Fig. 14.1 Gastronomy as a reason for visiting Dalmatia

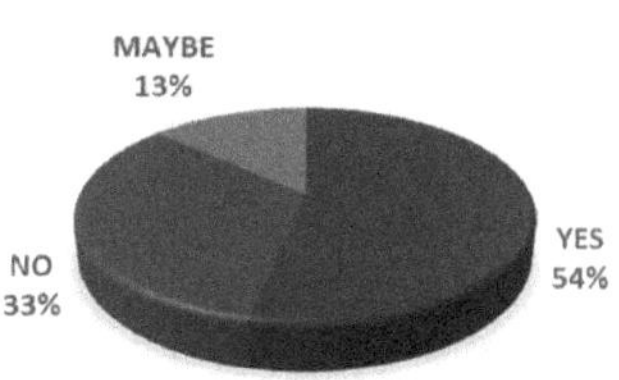

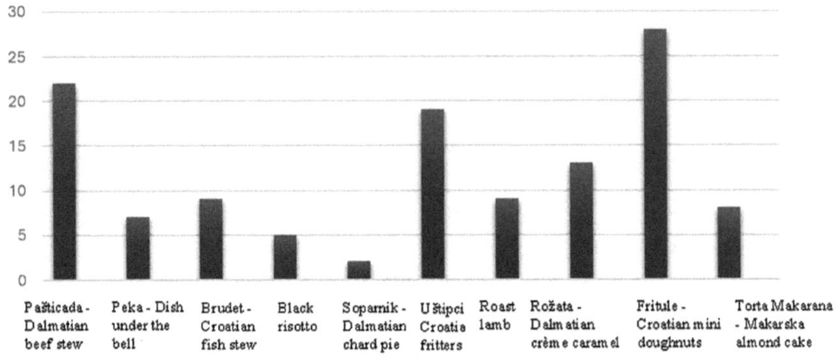

Fig. 14.2 Presentation of traditional dishes that guests tasted during their stay in Dalmatia

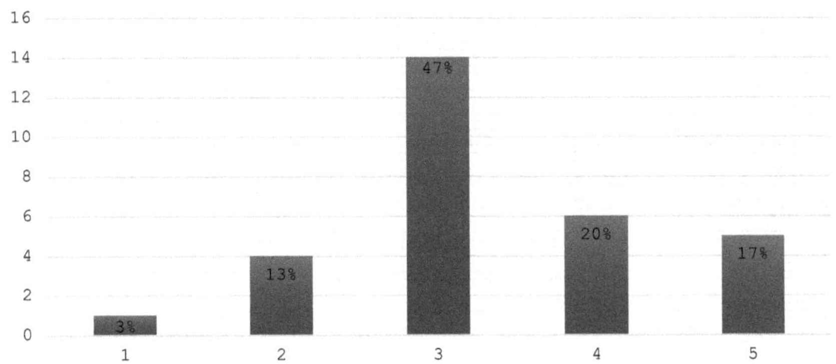

Fig. 14.3 Overall rating of Dalmatian cuisine

excellent. Only 13% of guests rated Dalmatian gastronomy as "passable," and a minor percentage of guests (3%) expressed a negative opinion.

Discussion

The sustainability of Dalmatian gastronomy and cuisine comes from the limited resources of the Mediterranean biome, which is reduced to only a few kilometers in some areas of Dalmatia. Under such conditions, the habit of exploiting the coast and sea as scarce resources has developed, accompanied by marked seasonality. The configuration of the Dalmatian coastline, with its numerous islands, closely resembles that of Greece. Additionally, when the first urban areas were created, essential olives, vines, and fig crops came to the Dalmatia coast and islands with ancient Greek settlers. This finding indicated that Dalmatian cuisine is very similar to the original type of Mediterranean diet (Calussi, 2008). The name Dalmatia comes from the Illyrian word for sheep, delma, from which we can conclude that sheep breeding has developed since ancient times in this region. Additionally, many toponyms indicate the breeding of goats. For example, the mountain Kozjak is a derivative of a goat, or the neighboring town Trogir, greek *tragus*, which translates into Capricorn. In the Mediterranean diet, milk is an important raw material obtained from these animals. The indented coast, with many islands and cliffs, was suitable for fishing and seafood as an important protein source, especially in late winter and early spring, when granaries were emptied and dried meat was eaten. The Jabuka Basin, which geographically belongs to Dalmatia, is the most affluent fishing area in the Adriatic Sea and an example of an efficient ecosystem of primary biomass production and a spawning place for small blue fish that serve as food for predatory species (Vitez & Muraj, 2001).

Therefore, it is not surprising that Adriatic has a thousand-year-old tradition of fishing and fish processing (Finka, 1995). Additionally, the location of Dalmatia on the waterway between Venice and Constantinople, in ancient times between Europe and the Levant, influenced Dalmatian cuisine to be enriched with spices, customs, and methods of preparing dishes from a much wider geographical area than itself. In addition, the continuity of life in the old Roman cities was maintained with many customs and recipes and in gastronomy. This is also supported by the fact that Petronius' "Satirikon," in which the feast of Trichmahlion is described, the most famous feast of imperial Rome, was found in Trogir, owned by the Cipiko family and is now kept in Paris (Lučin, 2014).

With the sea, beautiful coast, and nature, it is understandable that such a layered gastronomy attracts tourists and guests to Dalmatia. This research revealed that a high proportion (54%) of interviewed guests emphasized gastronomy (enjoying eating and drinking) as an important reason for visiting Dalmatia. These results are comparable to those of previous research from 2020 (39.2–69.7%), which depended on stratification according to the means of transport used to arrive at the destination and length of stay. The highest percentage of willingness to enjoy gastronomy is shown by guests who come with their car and stay at their destination for four to seven days (TOMAS, 2020).

Both studies showed that enjoying gastronomy is not related to the gender of the respondent. The results also revealed that most respondents tasted traditional Dalmatian dishes, corresponding to gastronomy, as the reason for going on independent or organized trips (59.2% in the previous survey). In the survey conducted, 84% of the guests rated Dalmatian cuisine as good to excellent, which is very close to the share of 80.5% who would recommend it to relatives, friends, and colleagues. Only 13% of guests rated Dalmatian gastronomy as satisfactory, and a small proportion (3%) expressed a negative opinion, which is close to the percentage of 12.7%, as noted by a survey conducted by the Institute for Tourism in 2019 (TOMAS, 2020).

Conclusion

Through its rich flavors and aromas and layering of historical periods and cultures, Dalmatian cuisine, food, and wine remain examples of a fundamental Mediterranean diet and important reasons for guests to visit this destination. Dalmatian gastronomy, with a wide variety of seafood and land, is an indispensable part of the hedonistic pleasure of every guest who visits Dalmatia. Considering the favorable climate and health benefits of the Mediterranean diet—alongside opportunities for active coastal holidays—it is unsurprising that Dalmatia's local population has greater vitality and a higher life expectancy than the national average (EUROSTAT, 2021). Today, in an era of increasing awareness of the importance of sustainable food production, the Mediterranean diet is gaining additional importance, as it mainly prefers food of plant origin, which contributes to reducing greenhouse gas emissions, using land, and using energy and water. Combining various positive factors with a willingness to adapt can encourage a shift toward a sustainable diet, one that respects both local gastronomy and the distinct elements of the food system (Fresán et al., 2018; Meybeck & Gitz, 2017).

References

Agnoli, C., Sieri, S., Ricceri, F., Giraudo, M. T., Masala, G., Assedi, M., Panico, S., Mattiello, A., Tumino, R., Giurdanella, M. C., & Krogh, V. (2018). Adherence to a Mediterranean diet and long-term changes in weight and waist circumference in the EPIC-Italy cohort. *Nutrition & Diabetes, 8*(1), 22. https://doi.org/10.1038/s41387-018-0023-3

Calussi, G. (1997). *Kuharica dalmatinskih gospođa*. Marjan tisak. Split.

Calussi, G. (2008). *Miris Dalmacije*. Marjan tisak. Split.

EUROSTAT (2021). Mortality and life expectancy statistics. Retrieved February 20, 2024, from https://ec.europa.eu/eurostat/statistics-explained/index.php?title=Online_publications

Finka, B. (1995). *Tisuću godina prvoga spomena ribarstva u Hrvata*. In: Kaštela, S. (Ed.) Tisuću godina prvoga spomena ribarstva u Hrvata (One Thousand Years Since the First Recorded Mention of Fishing Among the Croats)—Proceedings of a scientific conference. Croatian Academy of Sciences and Arts. Zagreb, Sali, Split.

Fresán, U., Martínez-Gonzalez, M. A., Sabaté, J., & Bes-Rastrollo, M. (2018). The Mediterranean diet, an environmentally friendly option: Evidence from the Seguimiento Universidad de Navarra (SUN) cohort. *Public Health Nutrition, 21*(8), 1573–1582. https://doi.org/10.1017/S1368980017003986

Lučin, B. (2014). *Petronije na istočnoj obali Jadrana Codex Traguriensis (Paris. Lat. 7989) i hrvatski humanisti*. Colloquia Maruliana XXIII. Split.

Marjanović Radica, D. (1976.). *Dalmatinska kuhinja*. Mladost. Zagreb.

Meybeck, A., & Gitz, V. (2017). Sustainable diets within sustainable food systems. *Proceedings of the Nutrition Society., 76*(1), 1–11. https://doi.org/10.1017/S0029665116000653

Montani, J. P. (2021). Ancel keys: The legacy of a giant in physiology, nutrition, and public health. *Obesity Reviews: An Official Journal of the International Association for the Study of Obesity, 22*(Suppl 2), e13196. https://doi.org/10.1111/obr.13196

TOMAS Hrvatska. (2020). *Stavovi i potrošnja turista u Hrvatskoj*. Institut za turizam. Zagreb.

Trichopoulou, A., & Vasilopoulou, E. (2000). Mediterranean diet and longevity. *The British Journal of Nutrition, 84*(Suppl 2), S205–S209. https://doi.org/10.1079/096582197388554

Tosti, V., Bertozzi, B., & Fontana, L. (2018). Health benefits of the mediterranean diet: Metabolic and molecular mechanisms. *The Journals of Gerontology. Series A, Biological Sciences and Medical Sciences, 73*(3), 318–326. https://doi.org/10.1093/gerona/glx227

Vitez, Z., & Muraj, A. (Eds.) (2001). *Hrvatska tradicijska kultura na razmeđu svjetova i epoha*. Galerija Klovićevi dvori - Institut za etnologiju i folkloristiku. Zagreb.

Zong, G., Gao, A., Hu, F. B., & Sun, Q. (2016). Whole grain intake and mortality from all causes, cardiovascular disease, and cancer: A meta-analysis of prospective cohort studies. *Circulation, 133*(24), 2370–2380. https://doi.org/10.1161/CIRCULATIONAHA.115.021101

CHAPTER 15

Catholic Ecology and the Contribution of the Franciscan Tradition to Integral Ecology

Luka Tomašević and Suzana Vuletić

Abstract Scientific ecology has garnered significant attention even in Christian circles and has initiated a new theological discipline: eco-theology or a new theology of creation. Climate change, desertification of land, destruction of the environment, and the emergence of uncontrolled development have prompted theologians to reexamine and speak out about the ordering of "God's house," that is, all that is created on Earth. The principle of human supremacy over all created things and all principles "inscribed in the human heart" was first called into question. Ecological thought has long matured within the Catholic Church. Finally,

L. Tomašević (Deceased)
Catholic Faculty of Theology, University of Split, Split, Croatia
e-mail: ltomasevic@kbf-st.hr

S. Vuletić (✉)
Catholic Faculty of Theology in Đakovo, Josip Juraj Strossmayer University of Osijek, Đakovo, Croatia
e-mail: suzanavuletic007@gmail.com

I. Pavić et al. (eds.), *Bioethics and Social Ethics in The Modern World*,
https://doi.org/10.1007/978-3-031-86418-6_15

Pope Francis's ecological encyclical appeared, profoundly engaging with ecological themes and the problem of the world's and man's development. Its ecological vision is holistic, similar to the vision of St. Francis of Assisi. In this sense, it forms the so-called third ecological path. Franciscan theology draws its roots from St. Francis and was most developed by St. Bonaventure and John Duns Scotus. In light of this, inspired by today's theological contributions, we point to the necessity of overcoming the prevailing technocratic paradigm according to which everything is directed toward the wonders of technology and finance and in which it is assumed that egoism is the primary driver of our system of values and actions. Currently, environmental issues are often not considered serious. However, they indicate how little attention contemporary society pays to them, although they are indicators of a profound ethical and anthropological crisis caused by human action. The Catholic Church has been providing solutions to the crisis since the very beginnings of ecology. Moreover, Pope Francis proposed and sought integral solutions inspired by Francis's theory. He noted how the human and social environment mutually deteriorates, while the social crisis is equated with the environmental crisis, and it cannot be separated from justice or from the human being itself. This chapter presents the moral crisis caused by the technocratic paradigm that fails to provide a comprehensive global vision. Owing to this fragmented conception, social relations are deteriorating; no one looks anyone in the eye, and everything is reduced to objects that are "used and thrown away." In the second part, some aspects of the Franciscan tradition are discussed, providing possible solutions to our severe socioenvironmental crisis.

Keywords Ecology · Theology · St. Francis · Franciscan school · Creation · God · God's plan · Church

The Moral Crisis of Society

The emergence of eco-theology and eco-philosophy (Deane-Drumond, 2008) and the advent of bioethics are necessarily linked to a significant moral crisis due to bio/ethical pluralism, which affects all of human life and its surroundings.

Owing to the biotechnological domination of nature, contemporary man has taken a despotic and dominant role above the whole created ambiance, not respecting its intrinsic legality of limited natural resources and the inability to recover. Biotechnology and the bioeconomy of modern industrial society are developing the world trade of natural resources on the basis of evaluations by the instrumental use of nature, led by their pragmatic interests, disregarding ethical imperatives. The consequence of man's aggression in the biosphere is environmental degradation and *"ecocide."* Since modern science, guided by pluralistic ethical paradigms, cannot provide a universally satisfactory answer to these challenges, multidisciplinary approaches, which range from bioethical ecological principles (Vuletić & Tomašević, 2014, 289).

In this atmosphere, amidst the confusion of practical ethics, ecology, and bioethics, it was born to respond to new applications regarding the world and humans (Poulmin, 1982) owing to a loss of respect for universal values.

In our era, there has been a loss of faith in philosophy, which no longer offers solutions to human problems, and a loss of faith in religious propositions and solutions. Even politics has changed in its purpose and mode of operation. A culture that focuses on the individual has developed, stimulating private autonomy, personal desires, consumption, and economic success. As a result, postmodern society lacks a system for general meaning or common good.

This is affecting the world, endangered by the scientific-technological approach of human supremacy over nature. Therefore, there is talk of the end of the scientific-technological epoch and the beginning of a new bioethical age.

However, the main weakness of this contemporary epochal turn and the demand for resolving the indicated problems is that we do not have a shared ethics or a typical resolution of concrete problems related to life and development.

Specifically, the Aristotelian-Thomistic ethics of purposefulness and reasonableness, along with all Christian ethical positions, have been rejected. All this has developed an individual morality for which no ordinary virtues exist. Therefore, man is turned exclusively to himself and believes that he is accountable only to his preferences. In this atmosphere, clear principles and firm stances have disappeared. They have been transformed into free personal options that affirm the individual but do not translate into general duties.

The crisis of the completion of the scientific-technological civilization, which crystallized precisely in the ethical implications of the way of life, can be resolved in the envisioned reversal of the mentioned epochs only on the basis of the spirituality from which this civilizational and world-historical cycle arose (Čović, 2004). In this context, the emergence of *eco-philosophy* and *eco-theology* needs to be considered.

The Emergence of Eco-Theology as a New Theological Discipline

Christianity is a religion of incarnation, embodying itself in every era, space, and culture, encountering the challenges of its time. The first reality of this embodied truth of life is the awareness that life in the world is harshly and painfully endangered by science and technology, so much that catastrophe threatens. Today, the Earth clearly shows its wounds (Aramini, 2010), whereas the renowned economist Serge Latouche claims that the most significant culprit is general development, which he calls the "source of evil" (Latouche, 2007). The second reality of Christianity's incarnation in the world is that it is a moral duty to preserve a healthy environment and improve and care for future generations who have a right to it (John Paul II, 1989). This leads us to eco-theology as a new theology that, for various reasons, has come into the spotlight of contemporary human interest and, in this sense, into anthropology and theology.

However, the ecological issue, as well as the clear awareness that technical science cannot give meaning to existence and life, forced theology to reconsider its anthropology and the relationship between humans and God among people and the relationship between humans and nature, leading to the mutual development of criticisms of Christianity and new theological reflections.

This movement gained extraordinary momentum and its mark from Saint Pope John Paul II with the sizeable ecological gathering in Assisi (1986), when he gathered world religions in prayer for peace and the preservation of creation, mainly through the development of millions of members and activists of various religious movements.

The World Council of Churches discussed Justice, Peace, and the Integrity of Creation in 1980. These years also saw the creation of ecological ethics under the influence of J. Moltmann and Catholic theologian M. Fox, which marked the beginning of the spirituality of deep ecology.

In the early 1990s, Earth Day was founded, and public messages for the New Year increasingly highlighted ecological ideas and eco-theology, whereas Leonardo Boff (1995) connected ecology with liberation theology. Today, there is also the World Forum on Theology and Liberation, which began in 1995 with the gathering under *"Ecology and Poverty: The Cry of the Earth, The Cry of the Poor"* (Concilium 5/1995). However, the Church's distinctly theological stance is found in the New Year's message for the *Day of Peace* in 1989, titled *"Peace with God the Creator—peace with all of creation."*

In short, theologians and religious groups in the 1980s opened new perspectives, and the 1990s saw many publications on these topics.

Theologians initially turned their thoughts to Teilhard de Chardin's (1976a, 1976b) ideas on evolution and creation, on the place of humans in nature and ideas that were deepened and expanded by a Catholic monk and theologian Thomas Berry (Berry, 2009) and philosopher Alfred North Whitehead (1996).

Through these meetings, theological reflections and research appeared and developed, leading to the emergence of eco-theology as a new branch of practical or constructive theology. From an ecological viewpoint, they proposed basic principles to emphasize the divine presence in the cosmos rather than a hierarchical understanding of God, humanity, and the created world. They reject the radical view of dominion and develop the idea of partnership with nature; it advocates for a vision of justice concerning creation, not only in the sphere of human needs but also in terms of all creation at a global level (Bouma-Prediger, 2001).

Eco-theology thus becomes part of constructive theology, focusing on the interrelationship between religion and nature, especially with respect to exploration. In general, it starts from the premise that there is a relationship between religion or the spiritual outlook and the degradation of nature. Therefore, its primary goal is the interaction between ecological values such as sustainability and human domination over nature. Two Jewish theologians and philosophers, Joshua Heschel and Martin Buber, significantly contributed to the development of this theology in the Christian world but also inspired Jewish eco-theology.

Similarly, the official Christian Church has been paying great attention to this ecological problem for decades. Pope John Paul II, Benedict XVI, and the current Pope Francis regularly address this issue.

Although the Catholic Church spoke about the ecological problem as early as 1971 in the Apostolic Letter of Pope Paul VI, *Octogesima adveniens*. The message at the conference in Stockholm was that the world we inherit is the heritage of all humanity and not just a privileged few. John Paul II, with his encyclicals *Solicitudo rei socialis* and *Centesimus annus* (n. 37.), brought full attention to ecological issues. He explicitly stated that at the root of the senseless destruction of the natural environment lies an anthropological error, which is unfortunately widespread in our time. Man, discovering his ability to transform and, in a certain sense, create the world through his work, forgets that this always takes place on the basis of the first original gift of things from God. Man thinks he can dispose of the Earth arbitrarily, subjecting it without restraint to his will, as if it did not have its form and a preestablished purpose given by God.

The main message is that environmental pollution is one of the dangerous aspects of the profound moral crisis of today's society. "*The ecological issue today is so vast that it includes the responsibility of all. Its various aspects require the necessity of coordinated efforts with a clear aim to define appropriate duties and tasks for individuals, nations, states, and the international community. Not only does this align with attempts to achieve true peace, but it also confirms and strengthens those attempts… There is an order in the world that must be respected. The human person, endowed with the ability to make free choices, has a clear responsibility to preserve this order, considering the well-being of future generations. The ecological crisis, I repeat, is a moral problem*" (John Paul II, 1989).

From these perspectives, humanity is faced with the moral necessity of new solidarity to reexamine its way of life. Therefore, the modern virtues of life should become modesty, moderation, self-discipline, and a spirit of sacrifice. This means educating people about their ecological responsibility and responsibility toward themselves, others, and the environment is necessary. This education cannot depend only on feelings and goodwill; it must not be merely an ideological or political orientation. "*True education for responsibility involves converting ways of thinking and behavior. In this regard, Churches and other religious institutions, government and nongovernmental organizations, indeed all stakeholders in society, must play their role.*" We need to establish a solidary. It is necessary to develop a sense of true brotherhood with everything surrounding us, as St. Francis of Assisi once emphasized.

Even Pope Benedict XVI dedicated significant attention to the ecological problem in his message for the 43rd World Day of Peace in 2010, titled *"If You Want to Cultivate Peace, Protect Creation"* (2009). He emphasizes that peace can be achieved through the defense and preservation of creation, a duty of our collective faithful and secular conscience of *"all men and women of good will."* He asserts that respect for creation is paramount because *"creation is the beginning and foundation of all God's works."* Thus, its preservation is essential for the peaceful coexistence of humanity, and all created things should be considered gifts to all humanity (Benedict XVI, 2009). On a practical level, he warns that *"the heritage of creation belongs to the entire human race,"* meaning that *"the current pace of exploitation is seriously endangering the availability of some natural resources not only for the present generation but especially for generations to come."* Therefore, it is not difficult to see that environmental destruction is often the result of a lack of long-term political projects or the myopic pursuit of economic interests, which unfortunately become a serious threat to everything created. It falls upon the international community and national governments. From these viewpoints, he encourages intergenerational and intragenerational solidarity. He warns everyone that environmental degradation calls for responsible behavior from producers and consumers, especially concerning basic ecological standards and energy resources.

Integral Ecology of Pope Francis

Pope Francis based his ecological ideas on the Franciscan concept of integral ecology, as elaborated in his encyclical *Laudato Si'* (2015). It begins with the words of the *Canticle of the Sun* by St. Francis of Assisi before addressing the damage humanity has inflicted on "our sister, Mother Earth," which, as a result, increasingly resembles *"an enormous pile of filth."* Similarly to his predecessors, Pope Francis emphasized the need for humanity to recognize the *"human roots of the ecological crisis"* and the *"anthropocentric excess."*

In the context of the current crisis, the Pope addresses all people of goodwill, calling for genuine ecological conversion and new universal solidarity. In this light, *Laudato Si'* directly speaks of obligatory care for our typical home and rationally warns that *"we are not God."* Earth precedes us and is *"given to us"* as a gift. For this reason, he cites St. Francis of Assisi, who was a "protector" and "witness" of what is now called integral ecology, enabling us to recognize in nature *"the magnificent book in which God speaks to us"* and where *"each creature has its value and purpose."*

St. Francis of Assisi is proclaimed the patron saint of ecology as an example for all Christians for his authentic and complete respect for the integrity of creation. Indeed, he is a model: *"The friend of the poor and beloved by God's creatures, he called all of them—animals, plants, natural forces, brother sun, and sister moon—to honor and praise the Lord. I wish his inspiration helps us always to maintain a lively sense of fraternity with all the good and beautiful things created by God Almighty and reminds us of our serious duty to respect and carefully preserve them within the broader and more significant human fraternity"* (John Paul II, 1989, 13). He understood the world as a place where we celebrate God and a house where we meet God. From these viewpoints, the world is our dwelling place, where we live and develop in a friendly, respectful, and loving manner.

This represents the so-called religious-sacramental stance of Christianity toward nature, which Pope Francis recognizes and accepts in his encyclical *"Laudato Si',"* proposing St. Francis as an example of *"integral ecology, lived in joy and authenticity"* (Pope Francis, 2020).

To promote an integral vision of eco-theology, it is necessary to create a normative system that includes insurmountable limitations and ensures the protection of ecosystems before new forms of power derived from technocracy end not only politics but also freedom and justice. It is necessary to undergo radical ecological conversion to achieve such goals.

References

Aramini, M. (2010). *La terra ferita. Etica e ambiente,* Saronno (Va).

Benedict XVI. (2009). *Ako želiš njegovati mir, čuvaj stvoreno,* Zagreb.

Berry, T. (2009). *The Christian future and the fate of earth.* Maryknoll.

Boff, L. (1995). *Sorella madre terra.* Roma.

Bouma-Predigerm, S. (2001). *For the beauty of the earth, a cristian vision for creation care.* Michigan.

Carbajo Nunez, M. (2017). *Ecologia francescana: Radici della Laudato Si'.*

Cobb, 1995Cobb, J. B. Jr. (1995). *Is it too late? A theology of ecology.* Environmental Ethics Books, Texas.

Čović, A. (2004). *Etika i bioetika. Razmišljanja na pragu bioetičke epohe.* Zagreb.

Deane-Drummond, C. (2008). *Eco-theology.*

Fox, M. W. (1996). *Agricide: the hidden farm and food crisis that affects us all.* Malabar.

Gottlieb, R. S. (2006). *A greener faith: Religious environmentalism and our planet's future*. https://doi.org/10.1093/acprof:oso/9780195176483.001.0001

Huntigton, S. P. (1997). *The clash of civilizations and the remaking of world order*.

John Paul II (1989). Mir s Bogom Stvoriteljem - mir s cjelokupnim stvorenjem. In: AKSA (22. 12. 1989.) (pp. 13–15).

John Paul II (1991, January 5). *Encyclical letter Centesimus Annus*. https://www.vatican.va/content/john-paul-ii/en/encyclicals/documents/hf_jp-ii_enc_01051991_centesimus-annus.html

Latouche, S. (2007). *Come sopravvivere allo sviluppo. Dalla decolonizzazione dell'immaginario economico alla costruzione di una societa' alternative*. Torino.

Moltmann, J. (1985). *Gott in der Schöpfung*. Ökologische Schöpfungslehre.

Pantheghini, G. (1992). *Il gemito della creazione. Ecologia e fede cristiana*.

Papinsko Vijeće "Iustitia et Pax" (2005). *Kompendij socijalnog nauka Crkve*.

Pavao, VI. (1991). Octogesima adveniens. In: M. Valković (Ed.) *Sto godina katoličkoga socijalnog nauka*. Zagreb.

Pope Francis, Encyclical letter Laudato Si' of the Holy Father Francis on Care for our Common Home, (24.5.2015.). Retrieved from https://www.vatican.va/content/francesco/en/encyclicals/documents/papa-francesco_20150524_enciclica-laudato-si.html

Pope Francis. (2020). *Fratelli tutti, Encyclical letter Fratelli Tutti of the Holy Father Francis on Fraternity and Social Friendship*, (3.10.2020.). Retrieved from https://www.vatican.va/content/francesco/en/encyclicals/documents/papa-francesco_20201003_enciclica-fratelli-tutti.html

Poulmin, S. (1982). How medicine saved the life of ethics. *Perspectives in Biology and Medicine*, *25*(4), 736–750. https://doi.org/10.1353/pbm.1982.0064

Pozaić, V. (1992). Teološki naglasci na skupu u Rio de Janeiru. *Socijalna ekologija*, *1*(3), 381–386.

Rogers, J. (1973), *Ecological theology. The search for an appropriate theological model*.

Teilhard de Chardin, P. (1969). *Comment je crois*. Paris (1933).

Teilhard de Chardin, P. (1976). *Il Posto dell'Uomo nella Natura-Il Gruppo zoologico umano*. Roma.

Vuletić, S., & Tomašević, L. (2014). Bioetički i eko-teološki poticaj za zaštitu okoliša i života. *Diacovensia*, *22*(3), 289–313.

Whitehead, N. (1996). *Wie entsteht Religion?*

CHAPTER 16

Environmental Sustainability in Times of Economic Crises: Ethical and Macroeconomic Dimensions

Lena Malešević Perović and Maja Mihaljević Kosor

Abstract This chapter contains an analysis of the impact of economic crises on environmental sustainability in the EU27 from 2000 to 2019, focusing on CO2 emissions, the material footprint, and natural resource depletion. The study reveals that economic crises significantly reduce these environmental indicators, highlighting a temporary respite from environmental degradation during downturns. Additionally, research has revealed a negative relationship between GDP and environmental indicators, which aligns with the environmental Kuznets curve theory. The findings underscore the potential for economic downturns to influence more sustainable consumption and production patterns in the short term. However, ethical considerations are essential, as economic crises may

L. Malešević Perović (✉) · M. Mihaljević Kosor
Faculty of Economics, Business and Tourism, University of Split, Split, Croatia
e-mail: lena@efst.hr

M. Mihaljević Kosor
e-mail: majam@efst.hr

I. Pavić et al. (eds.), *Bioethics and Social Ethics in The Modern World*,
https://doi.org/10.1007/978-3-031-86418-6_16

exacerbate social inequalities and vulnerabilities. The chapter emphasizes the need for policies prioritizing social equity, justice, and inclusivity in environmental interventions to achieve long-term environmental sustainability.

Keywords Economic crises · Sustainable development · Dynamic panels · Environment · SDGs

Introduction

Environmental sustainability becomes increasingly important as countries develop and achieve high growth rates. In this case, environmental indicators (such as air and water quality) tend to deteriorate (Cantone et al., 2021), urging heightened global environmental awareness. The need to act was recognized in 1992 when scientists issued the so-called "*World Scientists' Warning to Humanity*," emphasizing the danger of irreversibly damaging the environment (Union of Concerned Scientists, 1992). The United Nations (2015) 2030 Agenda for Sustainable Development and the European Commission's (2019) Green Deal represent two relevant initiatives to address global challenges and achieve sustainable and environmentally friendly economies. However, despite this recent heightened awareness, environmental issues often become overlooked and set aside during economic crises. Making ethical decisions about the environment in times of pronounced crisis and conflicting needs requires a better understanding of the ethical principles of sustainable development. At this point, sustainability represents an ethical framework for aligning economic development needs with environmental protection. We advocate that periods of economic crisis represent a significant risk for environmental objectives. Indeed, recent crises around the world (the COVID-19 crisis, wars in Ukraine and the Middle East, global inflation, and recession) have severely affected the achievement of sustainability (Leal Filho et al., 2023). Therefore, the goal of this chapter is to empirically assess the effects of economic crises on the environment in the EU27 during 2000–2019, taking into account the ethical dimensions of sustainability.

Ethical Dimensions of Sustainability

Sustainability encompasses three key dimensions: economic, social, and environmental. Economic ethics concerns moral foundations, characteristics, and consequences of economic activities and institutions. Economic ethics may also examine productivity, efficiency, and the role of technology. Economic ethics is sometimes considered a subfield of social ethics, but that is not the approach discussed in this chapter. Social ethics examines the ethical foundations and consequences of collective actions and decisions. It is concerned primarily with the goods that are collectively achieved. Environmental ethics examines entire ecosystems, regions, or smaller units, focusing on the value of nonhuman nature. Sustainability and the environment (ethics) are sometimes mistakenly used synonymously, although the environment is one of the three dimensions of sustainability. The outcome will be a healthy and strong society when these three dimensions are balanced. Given the previous discussion, it may be stated that, at its core, sustainability has ethical underpinnings. These ethical aspects are mostly implicit and consider utility, moral values, and goals. The needs of the population, both present and future generations, are fulfilled through a thoughtful assessment of resource consumption, economic development, technology, the role of government, and the protection of the environment (Kibert et al., 2011).

Making ethical decisions regarding the environment and economic development amid conflicting needs and approaches requires a clarification of the ethical principles surrounding sustainable development. The report published by the World Commission on Environment and Development (WCED, 1987), also known as the Brundtland Report, defines sustainable development as "development that meets the needs of the present without compromising the ability of future generations to meet their own needs" (WCED, p. 43). A more detailed discussion of the basis of this concept and the implications of the Brundtland Report can be found in Langhelle (1999), and various sustainable development definitions can be found in Pearce et al. (1989).

Historically, the term development was used to signify economic development, which was measured primarily through economic growth. Recent research has shown that long-term economic growth has positively affected sustainable development (e.g., Dima et al., 2024). Development ethics emerged as a part of a discussion on what development should encompass in addition to economic growth. Some of the first contributions in this field were from the 1960s, when Goulet stressed the importance of distinguishing between "authentic" human development and so-called "false" or "anti-development" (Goulet, 2006). For authentic human development, social and economic change is worth pursuing. It is a qualitative improvement in any society's provision of life-sustaining goods, esteem, and freedom to its citizens (Goulet, 1971).

Haq (1995) further advances the concept and discusses the four development goals. First, development must enhance people's well-being. Second, be equitable. Third, they should be empowered and, finally, environmentally sustainable. In addition, Drydyk (in Penz et al., 2011) added three more goals that enable the distinction between "authentic" and "false" development. These are human rights, cultural freedom, and integrity. A more comprehensive discussion of these seven human development goals can be found in Gasper (2004, 2012) and Drydyk and Keleher (2019). A valuable contribution to the discussion of ethical aspects and the dynamic character of the right to development can be found in Alston (2013). The concept of sustainability used here emphasizes the ethical approach that aligns the needs for economic development with environmental preservation and examines the consequences of their interplay.

Economic Crises and Environmental Sustainability

The connection between economic growth, a key macroeconomic variable, and the environment has been extensively explored in the literature. The so-called "environmental Kuznets curve" describes it (Grossman & Krueger, 1995). This inverted U-shaped curve accounts for the fact that low levels of GDP are associated with an increase in environmental deterioration, whereas higher levels of GDP lower this deterioration, i.e., lead to environmental improvement. A nonlinear shape, thus, best describes this initial positive and later negative impact on the environment. Initially, when countries are still underdeveloped and achieve low GDP levels, their development threatens the environment. There is more industrial production, more exhaustion of natural resources, more greenhouse gas emissions, and waste; therefore, the negative impact on the environment perseveres in this case. However, after countries reach a certain level of development (higher level of GDP), they tend to change the composition of their production toward greener solutions, care more about the environment, adopt policies that protect the environment, etc. Naturally, the relationship between GDP and environmental degradation becomes negative (Pacca et al., 2020).

The connection between economic crises and the environment has been far less investigated. Economic crises refer to periods of a decline in GDP. A fall in industrial production, which is a consequence of the crisis, then leads to a fall in investment, consumer, and energy consumption and consequently to less pollution (Jalles, 2020). Crises influence economic growth by definition; therefore, crises and GDP growth are closely related. However, these crises can pose a threat to the environment beyond the impact of GDP alone (Antonarakis et al., 2022; Antoniades et al., 2022; Malešević Perović & Ćorić, 2024). Moreover, crises also occur when environmental considerations tend to be put aside. Therefore, it is essential to investigate this relationship.

As indicated previously, the effects of economic crises on environmental sustainability have been underrepresented in the literature. Botetzagias et al. (2018) inspect the effect of economic crises on environmental performance in EU countries, taking into account whether a country has received the IMF/EU/ECB financial "rescue package" (the so-called "Troika effect"). When looking at the crises and the Troika effect separately, they find a nonsignificant to positive effect on several environmental quality indicators. However, the joint, i.e., the interaction effect of the two situations (crises + the rescue package), harms the environment. Pacca et al. (2020) explored the effects of financial crises on air pollutant emissions (CO2, SO2, NOx, and PM2.5). Their investigation suggested that, in the 1970–2014 period, air pollutant emissions initially decreased (except for PM2.5), followed by an increase or an insignificant effect in the medium term. This led them to conclude that the beneficial impact of financial crises on air quality is short-lived. Jalles (2020) examines the impact of various crises (financial, debt, banking, and currency) on air pollutant emissions. By investigating 55 countries from 1980 to 2012, he finds that banking crises negatively affect CO2 emissions, whereas debt and currency crises positively affect methane and fluorinated gas emissions. Antonarakis et al. (2022) explore a specific aspect of the environment—deforestation—and observe the impact of financial crises on this process. The results differ depending on the dataset used. The global Forest Watch results indicate that the overall effects are positive, i.e., the years of crisis were those of reduced deforestation. However, employing the global ESA-CCI forest cover dataset revealed a contrasting trend, indicating a global decrease in forest cover of 0.1 percentage points during financial crises.

Data, Methodology, and Results

For our empirical estimation of the described relationship between economic crises and the state of the environment, we apply the Blundell-Bond (1998) system GMM estimator. Following the approach of Antoniades et al. (2020) and Malešević Perović and Ćorić (2024), to maintain model parsimony, only economic crises and GDP per capita are employed as independent variables. The model is thus given as:

$$\text{Environment}_{\text{it}} = \alpha_i + \beta_1 \text{Environment}_{\text{it}-1} + \beta_2 \text{crises}_{\text{it}} + \beta_3 lnGDPpc_{\text{it}} + u_{\text{it}} \tag{16.1}$$

The environment is a measure of the state/quality of the environment and is, in our case, measured by three proxies used in turn. These are CO2 emissions per capita (in tons) (*CO2_em*), the material footprint per capita (in tons) (*mat_ftprint*), and natural resource depletion (% of GNI) (*nat_res_depl*). *Crises* are a dummy for economic crises in country i at time t, and lnGDPpc is the logarithm of GDP per capita. The error term, u_{it}, is a composed error.

We used a sample of 27 EU member states from 2000 to 2019 for further analysis. The definitions and sources of the variables used are provided in Table 16.1, while their descriptive statistics are presented in Table 16.2.

The results of the model, described by Eq. (16.1), are presented in Table 16.3., where each column refers to a different dependent variable, as explained previously.

The results in Table 16.3. suggest that the effect of economic crises on environmental indicators is statistically significant and negative. More precisely, we find that CO2 emissions (*CO2_em*) are minor during economic crises. This finding can be linked to the fact that at the beginning of the COVID-19 crisis, there was notable improvement in air quality, water cleanliness, and the overall state of nature. Specifically, as argued before, in times of crisis, there can be less energy consumption, industry production, and transport, which leads to less pollution and a reduction in CO2 emissions. Our results suggest that

Table 16.1 The definitions and sources of the variables

Variable	*Definition*	*Source*
Crises	Dummy for all economic crises (based on real GDP data), with the value of 1 in the first year of the crisis, 0 otherwise	Calculated by applying Harding and Pagan (2002) algorithm on the PWT10.0 NA data on real GDP in constant national prices
lnGDPpc	Logarithm of the expenditure-side real GDP at chained PPPs (in mil. 2017US$)	PWT10.0
CO2_em	CO2 emissions per capita (in tons)	Hickel (2020)
mat_ftprint	Material footprint per capita (in tons)	Hickel (2020)
nat_res_ depl	Adjusted savings: natural resources depletion (% of GNI)—Natural resource depletion is the sum of net forest depletion, energy depletion, and mineral depletion. Net forest depletion is unit resource rents times the excess of roundwood harvest over natural growth. Energy depletion is the ratio of the value of the stock of energy resources to the remaining reserve lifetime (capped at 25 years). It covers coal, crude oil, and natural gas. Mineral depletion is the ratio of the value of the stock of mineral resources to the remaining reserve lifetime (capped at 25 years). It covers tin, gold, lead, zinc, iron, copper, nickel, silver, bauxite, and phosphate	World Development Indicators

Table 16.2 Descriptive statistics of variables

Variable	*Obs*	*Mean*	*Std. dev*	*Min*	*Max*
Crises	520	0.14	0.35	0.00	1.00
lnGDPpc	520	10.42	0.42	9.01	11.63
CO2_em	500	9.01	2.48	3.84	14.92
mat_ftprint	500	23.86	6.31	6.80	39.93
nat_res_depl	494	0.24	0.37	0.00	2.16

economic crises reduce the material footprint, which refers to the utilization of all resources (renewable and nonrenewable) extracted to produce consumer goods and services. The MF can best be considered a measure reflecting the environmental consequences of economic activities. Our

Table 16.3 Results of the model

	(1) *CO2_em*	(2) *mat_ftprint*	(3) *nat_res_depl*
Crises	-0.636^{***}	-2.096^{***}	-0.0479^{***}
	(0.0585)	(0.145)	(0.0179)
lnGDPpc	-0.555^{***}	-1.186^{***}	-0.111^{***}
	(0.110)	(0.319)	(0.0345)
L.CO2_em	0.983^{***}		
	(0.0181)		
L.mat_ftprint		0.933^{***}	
		(0.0227)	
L.nat_res_depl			0.789^{***}
			(0.0218)
_cons	5.971^{***}	14.51^{***}	1.206^{***}
	(1.190)	(3.039)	(0.363)
N	475	475	469

Note: Standard errors are in parentheses
* $p < 0.10$, ** $p < 0.05$, *** $p < 0.01$

results suggest that economic crises reduce the use of these resources, thus reducing harmful environmental consequences and increasing environmental sustainability. The negative coefficient of the natural resource depletion variable suggests that times of economic crisis also occur when the depletion of natural resources declines. Since this indicator refers to various kinds of depletion (the sum of net forest depletion, energy depletion, and mineral depletion), it is a very positive outcome for the environment overall.

For GDP, we find the coefficient of this variable to be negative and statistically significant in all three cases. These results suggest that as a country's GDP increases, i.e., as it achieves a higher level of development and better living standards, CO2 emissions, the material footprint, and natural resource depletion tend to decline. This is in line with our previous discussion, whereby the environmental Kuznets curve suggested that countries at a higher level of development tend to care more about the environment. This finding makes sense since our sample consists of 27 EU countries, which all pertain to a group of more developed countries. Moreover, the statistically significant result for this variable, in addition to the crisis variable, lends support to the claim that economic crises affect the environment above the effect of GDP.

Finally, all lagged dependent variables (*L.CO2_em, L.mat_ftprint, L.nat_res_depl*) are statistically significant and positive, thus supporting the use and appropriateness of the dynamic model and suggesting that previous values of the environmental variables affect their current values.

Conclusion and Discussion

The empirical analysis undertaken in this chapter underscores the profound impact of economic crises on environmental outcomes. This suggests that during crisis periods, there was a noticeable reduction in environmental pressures, including CO2 emissions, the material footprint, and natural resource depletion. This phenomenon can be attributed to various factors, including decreased industrial activity, reduced energy consumption, and limited transportation, all of which contribute to mitigating environmental degradation.

The observed decline in CO2 emissions during economic crises reflects a reprieve from the relentless upward trend in greenhouse gas emissions typically associated with economic growth. This highlights the potential for significant environmental benefits when economic activities are scaled back or temporarily halted, as seen during lockdown measures implemented in response to the pandemic. However, it is essential to recognize that this reduction may not be sustainable in the long term and may rebound once economic activities resume.

Similarly, the decrease in the material footprint during economic crises signifies a reduction in resource extraction and consumption, which translates into lower environmental impacts associated with resource exploitation, production, and waste generation. This underscores the potential for economic downturns to facilitate transitions toward more sustainable consumption and production patterns, *albeit* temporarily.

Furthermore, the negative coefficient of natural resource depletion suggests that economic crises are conducive to reducing the exploitation of natural resources, including forests, energy sources, and minerals. This is a positive outcome for biodiversity conservation, ecosystem resilience, and the preservation of critical natural habitats, which are often threatened by extensive resource extraction during periods of economic growth.

However, it is crucial to acknowledge the ethical dimensions of environmental gains during economic crises. While reduced environmental impacts may seem beneficial from a sustainability perspective, economic downturns often exacerbate social inequalities and disproportionately affect vulnerable communities. Given that the literature on ethics and development has evolved over the past seventy years, it has been increasingly recognized that worthwhile development cannot be solely defined by economic growth. As previously discussed, several other goals must be achieved, such as enhancing people's well-being, ensuring equity and empowerment in sharing the benefits of environmentally sustainable development, and promoting human rights, culture, and integrity. Ethical considerations demand that environmental policies and interventions prioritize social equity, justice, and inclusivity, ensuring that marginalized populations do not disproportionately bear the burden of environmental protection and resilience-building efforts.

Moreover, the short-term nature of environmental gains during economic crises underscores the need for systemic changes prioritizing long-term sustainability over short-term economic gains. This requires transformative policies and interventions promoting sustainable development, environmental justice, and social resilience while addressing underlying structural inequalities and power imbalances perpetuating environmental degradation. Additional research, independent evaluations, and studies should lead to a better understanding of the consequences of economic crises on environmental sustainability and the ethical dilemmas they generate.

In summary, while economic crises may offer temporary respite from environmental degradation, achieving long-term environmental sustainability requires a holistic approach that integrates environmental, social, and ethical considerations into policy-making and decision-making processes. By prioritizing equity, justice, and inclusivity, societies can build resilience to future shocks and crises while fostering a more sustainable and equitable future for all.

References

Alston, P. (2013). The emergence of the right to development. In Office of the High Commissioner for Human Rights ed., *Realizing the Right to Development: Essays in Commemoration of 25 Years of the United Nations Declaration on the Right to Development*. United Nations, New York and Geneva, 7–16. https://doi.org/10.18356/2560d30c-en

Antonarakis, A. S., Pacca, L., & Antoniades, A. (2022). The effect of financial crises on deforestation: A global and regional panel data analysis. *Sustainability Science, 17*(3), 1037–1057. https://doi.org/10.1007/s11625-021-01086-8

Antoniades, A., Antonarakis, A. S., Gilman, J., & Kempf, I. (2022). *The crises-poverty-sustainability nexus in the context of the Sustainable Development Goals and Covid-19* (pp. 1–15). https://doi.org/10.1007/978-3-030-87417-9_1

Antoniades, A., Widiarto, I., & Antonarakis, A. S. (2020). Financial crises and the attainment of the SDGs: An adjusted multidimensional poverty approach. *Sustainability Science, 15*(6), 1683–1698. https://doi.org/10.1007/s11625-019-00771-z

Blundell, R., & Bond, B. (1998). Initial conditions and moment restrictions in dynamic panel data models. *Journal of Econometrics, 87*, 115–143. https://doi.org/10.1016/S0304-4076(98)00009-8

Botetzagias, I., Tsagkari, M., & Malesios, C. (2018). Is the 'Troika' bad for the environment? An analysis of EU countries' environmental performance in times of economic downturn and austerity memoranda. *Ecological Economics, 150*, 34–51. https://doi.org/10.1016/j.ecolecon.2018.04.001

Cantone, B., Antonarakis, A. S., & Antoniades, A. (2021). The great stagnation and environmental sustainability: A multidimensional perspective. *Sustainable Development., 29*, 485–503. https://doi.org/10.1002/sd.2195

Dima, B., Dima, Ş. M., & Tudor, A. T. (2024). Societal sustainable development and long-run economic growth: How do we stand? *Sustainable Development*, 1–28. https://doi.org/10.1002/sd.2896

Drydyk, J., & Keleher, L. (Eds.). (2019). *Routledge handbook of development ethics* (1st ed.). Routledge. https://doi.org/10.4324/9781315626796-1

European Commission. (2019). European Green Deal. Retrieved May 5, 2023, from https://ec.europa.eu/info/strategy/priorities-2019-2024/european-green-deal_en

Gasper, D. (2004). The ethics of development: From economism to human development. *Edinburgh University Press*. https://doi.org/10.1080/17449626.2012.672450

Gasper, D. (2012). Development ethics—Why? What? How? A formulation of the field. *Journal of Global Ethics, 8*(1), 117–135.

Goulet, D. (1971). *The cruel choice: A new concept in the theory of development*. Atheneum. https://doi.org/10.4324/9780203086643

Goulet, D. (2006). *Development ethics at work: Explorations 1960–2002*. Routledge.

Grossman, G. M., & Krueger, A. B. (1995). Economic growth and the environment. *In Source: The Quarterly Journal of Economics, 110*, 353–377. https://doi.org/10.2307/2118443

Haq, M. (1995). *Reflections on human development*. New York: Oxford University Press.https://doi.org/10.1093/oso/9780195101911.001.0001

Harding, D., & Pagan, A. (2002). Dissecting the cycle: A methodological investigation. *Journal of Monetary Economics, 49*, 365–381. https://doi.org/10.1016/S0304-3932(01)00108-8

Hickel, J. (2020). The sustainable development index: Measuring the ecological efficiency of human development in the anthropocene. *Ecological Economics, 167*. https://doi.org/10.1016/j.ecolecon.2019.05.011

Jalles, J. T. (2020). The impact of financial crises on the environment in developing countries. *Annals of Finance, 16*(2), 281–306. https://doi.org/10.1007/s10436-019-00356-x

Langhelle, O. (1999). Sustainable development: Exploring the ethics of our common future. *International Political Science Review, 20*(2), 129–149. https://doi.org/10.1177/0192512199202002

Leal Filho, W., Trevisan, L. V., Rampasso, I. S., Anholon, R., Dinis, M. A. P., Brandli, L. L., Sierra, J., Salvia, A. L., Pretorius, R., Nicolau, M., Eustachio, J. P. P., & Mazutti, J. (2023). When the alarm bells ring: Why the UN sustainable development goals may not be achieved by 2030. *Journal of Cleaner Production, 407*, 137108. https://doi.org/10.1016/j.jclepro.2023.137108

Kibert, C. J., Peterson, A., Thiele, L., Monroe, M. & Plate, R. (2011). *Working toward sustainability: Ethical decision-making in a technological world* (1st ed.). Wiley.

Malešević Perović, L., & Ćorić, B. (2024). Sustainable development and economic disasters. *Journal of Cleaner Production, 434*, 140043. https://doi.org/10.1016/j.jclepro.2023.140043

Pacca, L., Antonarakis, A., Schröder, P., & Antoniades, A. (2020). The effect of financial crises on air pollutant emissions: An assessment of the short vs. medium-term effects. *Science of the Total Environment, 698*. https://doi.org/10.1016/j.scitotenv.2019.133614

Pearce, D. W., Markandya, A., & Barbier, E. B. (1989). *Blueprint for a green economy*. https://doi.org/10.4324/9780203097298

Penz, P., Drydyk, J., & Bose, P. S. (2011). *Displacement by development: Ethics, rights and responsibilities*. Cambridge University Press. https://doi.org/10.1017/CBO9780511973499

Union of Concerned Scientists. (1992). World Scientists' Warning to Humanity. Union of Concerned Scientists.

United Nations. (2015). *Transforming our world: The 2030 agenda for sustainable development*. Resolution Adopted by the General Assembly on 25 September 2015, 42809, 1–13.

WCED (The World Commission on Environment and Development). (1987). *Our Common Future*. Oxford University Press.

World Development Indicators. Washington, D.C.: The World Bank.

CHAPTER 17

Religious Tourism and the Environment: World Youth Day 2023 and Pope Francis' Environmental Catechesis

Edward R. Grant and Marija Barić Đurđević

Abstract The Catholic Church, the world's largest religious organization, has actively participated in discussions on environmental protection during the past fifty years. The three most recent popes, John Paul II, Benedict XVI, and Francis, all issued primary teaching documents (encyclicals) on the subject, culminating in "Laudato Si," the 2015 encyclical by Pope Francis that cemented protection of the environment as a core component of Catholic social teaching. World Youth Day, founded by John Paul II, is the world's largest religious gathering for young people. World Youth Day edition in Lisbon placed special emphasis on

E. R. Grant
Georgetown University, Washington, DC, USA
e-mail: Erg24@georgetown.edu

M. Barić Đurđević (✉)
University of Split, Split, Croatia
e-mail: marija.baric.durdevic2@gmail.com

I. Pavić et al. (eds.), *Bioethics and Social Ethics in The Modern World*,
https://doi.org/10.1007/978-3-031-86418-6_17

environmental protection. The official program, which consisted of spiritual, historical, and tourist content, gave the pilgrims clear instructions on environmental protection. While many of the tips, such as advice on waste recycling, were met with wide approval, some requirements, which asked for greater sacrifice, met strong resistance. In this paper, we look at the contributions of the Catholic Church and recent popes to contemporary environmental issues. Laudato Si covers many scientific, political, cultural, and theological issues related to the environment. In keeping with his predecessors, Pope Francis emphasized that in addressing such issues, we must realize that everything in the created world is connected, that we must move away from those aspects of the "technological imperative" that tend to dissolve this connectivity and that, as a result, lifestyles (particularly in the developed world) must change. The last injunction is widely shared by those (secular or religious) concerned with environmental protection. However, in keeping with the injunction that "everything is connected," Francis (as well as John Paul and Benedict) proposed a lifestyle change that is not widely accepted. The dominant feature of Western lifestyles in the past two generations has been individualism, which is characterized by sexual liberation, a decrease in families, and a decline in religion. We greatly privilege this brand of lifestyle liberalism and individualism. Nevertheless, as "Laudato Si" emphasized, its rise has led to interior emptiness and social fragmentation. To compensate, we desire more material things, innovations, and experiences—the desires that feed the dynamics of materialism and consumerism. In Pope's view, any effective lifestyle change must address the resulting spiritual weakness that denies the ultimate connectivity of all created things. Those concerned with environmentalism, particularly climate change, are happy with Pope's support. However, are they willing to turn away from these aspects of the modern Western lifestyle that (in concert with other factors) create the consumerist demand that the Pope urges us to abandon? World Youth Day was the first universal event sponsored by the Church, during which we could see Laudato Si in practice. How did it go?

Keywords Religious tourism · Laudato Si · Catholic Church · Pope Francis · World Youth Day · Environmental protection · Technological imperative · Individualism · Materialism

Introduction

Travel, in the form of pilgrimage, has been a hallmark of Christian piety for many centuries. This should not be confused with mere tourism—although pilgrims can become tourists, and vice versa, in the modern age of global travel and flexible itineraries. Both in the past and present, pilgrims travel to places associated with Saints or, in the case of the Holy Land, Jesus Christ himself. In addition to Jerusalem and other Holy Land sites, one thinks of Rome and Vatican City, Lourdes in France, Santiago de Compostela in Spain, and, through the Middle Ages, the pilgrimage to the shrine of St. Thomas Becket, immortalized by Geoffrey Chaucer in "Canterbury Tales."

In recent decades, a new form of "pilgrimage" has emerged among the Catholic youth of the world. First convened in 1995 by Pope St. John Paul II, the 17th celebration of "World Youth Day" (WYD) took place in August of 2024 in Lisbon, Portugal. The World Youth Day is the world's most prominent recurring youth event, according to Singleton (2011, 32). While some editions of World Youth Day have occurred in traditional places of pilgrimage, such as Rome, Czestochowa in Poland, and Santiago de Compostella, many have occurred in large, secular, or nonobservant cities: Paris, Toronto, Sydney, Cologne, and Denver. The 2027 WYD is scheduled for Seoul, South Korea—a tribute to the fortitude of the nation's historically persecuted Christian minority but hardly a place of traditional Catholic pilgrimage.

Nevertheless, the spirit of pilgrimage remains. Many young people walk a considerable distance to the pilgrimage site. Once there, living conditions are modest, tent cities abound, and overnight vigils, even if not observed in cloistered silence, are common. As Pope Benedict stated in his homily at Sydney WYD in 2008, the joy expressed at these events is not the ecstatic expression of a rock concert but is derived from the presence of the Holy Spirit (Benedict XVI, 2008).

Pope Francis presided over his fourth WYD. The official theme was "*Mary arose and went with haste—the bridge between the accounts, in the Gospel of Luke, of Gabriel's Annunciation of the Incarnation and Mary's Visitation to her cousin Elizabeth*" (Francis, 2022). Nevertheless, on the ground, there was another pervasive theme: carrying out in practical ways the "environmental catechesis" of Pope Francis, pronounced most authoritatively in the lengthy 2015 encyclical, *Laudato Si*—in English, "Praise be," and in Croatian "Hvaljen budi." This encyclical, as Joshtrom

Isaac Kureethadam (2016) says, is ground-breaking; it is the Pope's call to all the people of goodwill to actively evaluate their impact on the environment and current questions of climate change. Written in Italian instead of the traditional Latin of official Vatican documents, it was addressed to the entire world, believers and nonbelievers alike.

A Call for Lifestiyle Change: The Fight with Consumerism

Working with their cosponsors in the national and municipal governments, WYD organizers arranged to plant thousands of trees to offset the carbon footprint of such a vast event. Plastic and paper were minimized: meals were served in uniform containers designed for prompt recycling; the official program, which was replete with historical, spiritual, and tourist content, was exclusively online. More controversially, the hundreds of thousands of participants who traveled by air were asked to consider foregoing meat for a year to offset the resulting carbon emissions.

However, these practical measures—which can be found at any secular music festival—are insufficient to meet the demands of the conversion called for in *Laudato Si*. The lengthy encyclical, directed not only to Catholics but also to the entire world, covers many scientific, political, cultural, and theological issues related to the environment—far too much to summarize here. Its overriding themes echo the thought of both John Paul II and Benedict XVI: first, in addressing environmental issues, we must realize that everything in the created world is connected; second, we must move away from those aspects of the "technological paradigm" that tend to fracture this connectivity; and third, lifestyles (particularly in the developed world) consequently must change.

This last injunction, regarding the simplification of lifestyles, is widely shared, sometimes more in conviction than in practice, by those most vocal in their concern for the environment and the effects of climate change. *Laudato Si* is explicit, quoting Pope Benedict, who wrote in his 2010 message for the World Day of Peace: *"technologically advanced societies must be prepared to encourage more sober lifestyles while reducing their energy consumption and improving its efficiency"* (Benedict XVI, 2010). Pope Francis, while advocating large-scale economic and political reforms beyond the scope of these remarks, lauds those whose convert, in small ways, move in this direction. *"There is a nobility in the duty to care for the environment in little daily actions, and it is wonderful how education can bring about real lifestyle changes"* (Francis, 2015).

However, in keeping with the understanding that "everything is connected," Francis, following his predecessors, enjoins lifestyle changes that will not be widely accepted by the secular, materialist societies of the "West" and "North." The "consumerist" mentality rampant in these societies is not limited to acquiring material goods. Instead, it extends to a form of "lifestyle libertarianism"—even "libertinism"—that has led to severe declines in religious practices, family formation, and birth rates. Contrary to many voices in secular ecological movements, Pope Francis is adamant that "demographic growth is fully compatible with an integral and shared development" of humanity. To blame population growth in poorer countries rather than *"extreme and excessive consumerism" in* wealthier countries, he adds to avoid this issue (Francis, 2015).

While Francis did not explicitly state this, it is a simple fact that the rapid growth in materialism in the world's advanced economies has coincided with precipitous declines in the birth rate and family formation. The availability of disposable income that otherwise might be devoted to raising children feeds the cycle of consumerism. However, as Francis noted, the difficulty of "rising up" to the challenge of ecological conversion *"has much to do with an ethical and cultural decline that has accompanied environmental deterioration. Men and women of our postmodern world run the risk of rampant individualism, and many problems of society are connected with today's self-centered culture of instant gratification"* (Francis, 2015).

Many aspects of modern tourism reflect this. There is a drive to collect experiences, sometimes more than material goods. The plight of Venice offers a poignant example. Overtourism, exacerbated by the arrival of many cruise ships disembarking thousands of passengers at a time, threatens the jewel of Adriatic as a living city. Another extreme is the scandal of "mountaineering tourism" at peaks such as Mount Everest. Amateur adventurers with deep pockets pay exorbitant amounts to be able to claim that they stood on the "top of the world"—and leave in their wake mountains of discarded gear and, tragically, the bodies of those who, in all honesty, should never have attempted. Even the Galapagos Islands, which are treasured for their unique biodiversity, are threatened by "ecotourism."

Distinguishing between legitimate and admirable exploration of the Earth's surface and oceans and acquisitive adventurism can be challenging. Part of the answer lies in applying the highest of the "natural" virtues—prudence. However, the ultimate answer will require a more profound conversion. As Pope Francis explains, the acquisitive drive present in modern society is often fed by spiritual emptiness or dissatisfaction. In keeping with the theme of interconnectedness, this spiritual deficit and the related challenge of humankind's relationship to the natural endowment given to us by our Creator are rooted in the concept of human dignity. To reject our "createdness" as rational creatures with responsibility for stewardship over the rest of Creation is an affront to that inherent dignity and diminishes our worth. Francis extended this invitation: *"Those who are committed to defending human dignity can find in the Christian faith the deepest reasons for this commitment. How wonderful is the certainty that each human life is not adrift in the midst of hopeless chaos in a world ruled by pure chance or endlessly recurring cycles!"* (Francis, 2015).

The most profound question is whether the world will consider this invitation. Technological fixes—including efforts to create "clean" energy, convert it to electric vehicles, etc., will not solve the deficits in "human ecology" that are at the root of our environmental challenges. How well, in retrospect, did WYD 2023 do in calling the world's youth to accept, with joy, this invitation?

Harmonizing Care for Enviorment and Spiritual Experience at World Youth Day: Participant Reactions and Controversies

Several practical measures undertaken by WYD organizers have already been mentioned. One of the organizers, Pedro Madureira, told an American interviewer that these measures were more than mere window dressing and included how the WYD office met its supply needs. *"I do not think we bought anything for the offices,"* he said. *"Everything was either donated or remodeled. There are no single-use cutlery and dishes; we bring everything from home, so we don't throw it away, or we wash it ourselves"*(D'Avillez, 31.7.2023). While these and similar measures to ameliorate environmental impact before event were generally approved, some brought controversy, such as the recommendation to forego meat.

Portuguese farmers and small meat producers sent a formal protest note; medical profession members also objected, reminding attendees of the possible adverse health consequences of such a diet. Furthermore, widespread criticism, and some derision, greeted the news that the official WYD "app" included a carbon footprint calculator, purportedly allowing pilgrims to see how much their participation contributed to global warming. Francisco Correia, a university law professor in Portugal who hosted some WYD pilgrims, voiced concern that these measures involved the Church too deeply in political matters (D'Avillez, 2023a, 2023b).

Nevertheless, the most important assessment must come from the participants themselves. We spoke with three young attendees, hailing from Belgium, Croatia, and the United States, ranging in age from 26 to 28. The interviewees were asked whether, in their view, the event met its goals of being focused on ecological concerns. All three answered affirmatively. As one example, they noted that the backpacks they received included entirely reusable materials for food and drinks and that the use of paper was reduced to a minimum. While most of Pope's sermons focused on spiritual remedies for the issues of anxiety and depression that affect so many young people, the non-liturgical program was filled with workshops, film screenings, and activities with the theme of environmental protection. (Francis 2023b)

Some efforts to inculcate environmental themes were less successful than others. A Croatian man said that he participated in a workshop (attendance at workshops was assigned, not at discretion) in which lyrics of the song "Let It Go" (from the Disney movie Frozen) were altered to send a message about environmental protection. When he felt uncomfortable at this point, he left the workshop. The three interviewed participants noted that the heavy focus on environmental issues led to a clash of ideals. Specifically, they wondered if environmental concerns had replaced God in this iteration of WYD. Is the "green movement" so strong that it has become a preoccupation of the Church? Young people came for a spiritual experience and did not receive that to the extent they expected.

Certain—perhaps many—aspects of this "green movement" are ideological and thus do not map well into the genuine Christian faith. The more extreme aspects of this ideology are evidenced by numerous aggressive protests, such as the attempted destruction of Van Gogh's artworks, paintings in the Louvre and the National Gallery, and the dangerous blocking of roads and highways. Environmental protection is an omnipresent topic in public opinion but is often approached with fearful rhetoric. The laudable goal of enlightening all humanity to care for our typical home is ill-served by a posture of fear or by dogmatic and ideological approaches. It is unclear whether WYD 2023 did enough to inspire an approach to environmental issues deeply rooted in a Christian view of stewardship of God's gift of Creation. This view should never be confused with or corrupted by ideology.

Sydney Archbishop Anthony Fisher, O.P., expressed the danger of this corruption in remarks delivered shortly before the 2023 meeting of the Synod on Synodality (Fisher, 2023).

Conclusion

The Pope reminds us that the environment is one of the more significant concerns. During Synod's first week, the Holy Father issued *Laudate Deum*, his addendum to *Laudato Si*. In both documents, he challenges attitudes and behaviors that reduce our typical home, the Earth, our plaything, to exploit and damage at will in service of our interests and ideologies. He recalls that the world is given into our hands as a sacred trust to be reverenced, shared, developed, and handed on intact to future generations. (Francis, 2023a) Therefore, it is suggested that we must challenge attitudes to our typical home, the Church—refusing to treat it as our plaything, subservient to our interests and ideologies, to be remade at will. With her Gospel and mission, the Church is the new Creation, a spiritual ecosystem, given into our hands by the Lord Jesus as a sacred trust, to be reverenced and shared, developed and handed on intact to future generations.

Infused by ideology, the world outside the Church too often approaches the problem of environmental degradation in the wrong way. If we become convulsively afraid of the future, environmental protection becomes a bitter chore rather than a missionary endeavor of Christian witnesses. A feeling of guilt will prevail, and many will give up because they will not see the point of so much effort when the future is so dark.

It is necessary to change the view and approach. The Catholic Church, as a longstanding arbiter of moral guidance, offers answers to this crisis, just as it has offered its approach to every problem until now. In this context, the initiator should have a sense of admiration and gratitude instead of an extreme sense of duty. The exclamation "Laudato Si, mi Signore" is precisely that—an expression of gratitude and admiration by

St. Francis toward the Creator. As Kureethadam (2016) noted, admiration and gratitude can only come from knowing God's goodness, greatness, and genius. Given the fear of God, the person realizes not only its smallness but also thc privilege of participating in inhabiting this magnificent habitant. If we saw ingenuity and miracles in everything created, delight and awe would compel us to care for our environment. People would fear lest we offend the Creator, but not in the fear that an enslaved person has for his master. However, as a young man, he does not want to offend a girl he loves; everything about her is beautiful. He, therefore, does everything for her, respects her, and takes care of everything that belongs to her. Of course, the girl does not remain in debt. She returns the young man's gratitude for her beauty, lights up his life, and returns his respect.

> *Just as happens when we fall in love with someone, whenever he would gaze at the sun, the moon or the smallest of animals, he burst into song, drawing all other creatures into his praise. He communed with all Creation, even preaching to the flowers, inviting them "to praise the Lord, just as if they were endowed with reason."* (Francis 2015)

From a Catholic perspective, to achieve admiration, a person must overcome arrogance and admit that there is something more than himself. That person is only a creature, and above him is the Creator. Man must first build his relationship with God, which is a long-term process. It is impossible to overcome these problems with only material solutions but with the full capacity of human genius. This approach, which is based on admiration and gratitude as the central postulate, has no fear like the first approach does and is more long term. Such an approach lies in the fundaments of philosophy since Plato itself reminds us of the intrinsic alignment between aesthetic beauty and moral goodness.

References

Benedict XVI. (2008). *Homily of His Holiness Benedict XVI at the Eucharistic Celebration of the XXIII World Youth Day, 20 July 2008*. Retrieved April 19, 2024, from https://www.vatican.va/content/benedict-xvi/en/homilies/2008/documents/hf_ben-xvi_hom_20080720_xxiii-wyd.html

Benedict XVI. (2010). Message for World Day of Peace for year 2010. Retrieved April 19, 2024, from https://www.vatican.va/content/benedict-xvi/en/messages/peace/documents/hf_ben-xvi_mes_20091208_xliii-world-day-peace.html

D'Avillez, F. (2023a). At World Youth Day, it's not easy being green. *The Pillar*, 31 July. Retrieved April 19, 2024, from https://www.pillarcatholic.com/p/at-world-youth-day-its-not-easy-being

D'Avillez, F. (2023b). At World Youth Day, The biggest job I'll ever have. *The Pillar,* 4. August. Retrieved April 19, 2024, from https://www.pillarcatholic.com/p/at-world-youth-day-the-biggest-job

Fisher, A. (2023). *Walking together in communion, participation and mission. Reflections on the Synod on Synodality.* Retrieved April 19, 2024, from https://www.sydneycatholic.org/addresses-and-statements/2023/walking-together-in-communion-participation-and-mission-reflections-on-the-synod-on-synodality/

Francis. (2015). *Encyclical letter "Laudatio si" on care for our common home.* Retrieved April 19, 2024, from https://www.vatican.va/content/francesco/en/encyclicals/documents/papa-francesco_20150524_enciclica-laudato-si.html

Francis. (2022). *Message of his holiness Pope Francis for the XXXVII World Youth Day 2022–2023.* Retrieved April 19, 2024, from https://www.vatican.va/content/francesco/en/messages/youth/documents/papa-francesco_20220815_messaggio-giovani_2022.html

Francis. (2023a). *Apostolic exhortation "Laudate Deum."* Retrieved April 20, 2024, from https://www.vatican.va/content/francesco/en/apost_exhortations/documents/20231004-laudate-deum.html

Francis. (2023b). *Address of his holiness at the world youth day welcome ceremony.* Retrieved April 20, 2024, from https://www.vatican.va/content/francesco/en/speeches/2023/august/documents/20230803-portogallo-cerimonia-accoglienza.html

History of World Youth Day. Retrieved April 20, 2024, from https://worldyouthday.com/about-wyd/wyd-history/

Kureethadam, J. I. (2016). Ecological virtues in Laudato Si, *Ethics In Progress, 7*(1), 44–66. Retrieved April 19, 2024, from https://doi.org/10.14746/eip.2016.1.4

Singleton, A. (2011). The impact of World Youth Day on religious practice. *Journal of Beliefs & Values, 32*(1)

CHAPTER 18

Medical and Social Risks of Health Tourism

Silvana Pavlinović, Suzana Vuletić, and Gordana Pelčić

Abstract The health tourism industry is rapidly expanding. Clients often travel abroad for medically motivated trips due to preventing or rehabilitating illness, treating acute diseases, mitigating chronic disease consequences, or maintaining good psycho-physical conditions through recreational wellness, spas, and bath offers. Other reasons include lower costs of treatments abroad, shorter waiting lists compared with their home country, availability of advanced biomedical innovations, affordability of some medical procedures, or less stringent regulations than in their home country. Medical tourism has developed primarily for

S. Pavlinović
University of Split, Study of Hotel Management and Gastronomy in Makarska, Makarska, Croatia
e-mail: hotel@hotel-svetikriz.hr

S. Vuletić (✉)
Catholic Faculty of Theology in Đakovo, Josip Juraj Strossmayer University of Osijek, Đakovo, Croatia
e-mail: suzanavuletic007@gmail.com

G. Pelčić
Faculty of Medicine, University of Rijeka, Rijeka, Croatia
e-mail: gordana.pelcic@medri.uniri.hr

I. Pavić et al. (eds.), *Bioethics and Social Ethics in The Modern World*,
https://doi.org/10.1007/978-3-031-86418-6_18

two fundamental reasons: economic accessibility and permissible bioethical legislation regarding some clinical issues. However, many areas of medical tourism make space for questionable medical quality standards of requested procedures or (un)permitted health services, ignoring the bioethical principle of equal access to healthcare services and many other questionable impacts on health. This is why health tourism demands a clear need for a precise definition of unified quality standards and bioethical norms in health protection at the international level.

Keywords Health tourism · Well-being · Commercialization · Quality of medical service · Health protection · Bioethics

Introduction

Health tourism, encompassing medical-recreational vacations, is increasingly vital in revitalizing individuals' psycho-physical well-being. With growing awareness of preventive and rehabilitative health maintenance and a holistic well-being orientation, health tourism has become a primary travel motivation to destinations with medicinal natural resources.

The medical tourism index is underpinned by three principal dimensions: the broader national context (encompassing economic conditions, safety, image, and culture), the medical tourism industry itself (addressing destination appeal and cost), and the quality of facilities and services (encompassing clinical standards, reputation, global reach, accreditation, and the overall patient experience). This combined recreational vacation and medical service model, often including tourist and gastronomic experiences, attracts individuals for various reasons.

The primary motivation for traveling for medical treatment is obtaining superior healthcare at a lower cost. Other reasons include the inability to undergo specific medical interventions in their home country, better availability of state-of-the-art medical equipment, superior postintervention care, or faster procedures due to long waiting lists in the country of residence (Han & Hyun, 2015).

Health tourism often involves specialized hotels with integrated facilities of smaller ordinations, polyclinics, or specific hospitals, primarily for dentistry, aesthetic and plastic surgery, dermatology, ophthalmology, physical medicine and rehabilitation, infertility treatment, etc.

However, globalization trends have also introduced the lucrative aspect of medical tourism, which conflicts with bioethical principles and legal regulations between the domicile and the recipient country. This includes issues such as black markets for organ transplantation, regenerative medicine using embryonic stem cells, surrogacy, euthanasia, abortion, cryopreservation, and cyborg medicine. Given these moral concerns, this article aims to highlight such bioethical issues and analyze the motivational reasons for the international demand for health tourism services and their legal consequences.

Conceptual Variations in Health Tourism

Health tourism lacks a universally accepted definition, varying depending on the authors (Menvielle et al., 2011; Müller & Kauffman, 2001). Generally, it is defined as organized travel outside one's residence to improve mental, physical, and emotional well-being; promote health; and help one recover from illness (Carrera & Bridges, 2006). It aims to treat or improve health via the use of natural environmental factors (Špoljar, 1994, 163).

The Republic of Croatia's *National Strategy for Health Development* (2012–2020, *ad* 3.12.) describes health tourism through the following concepts:

- *Tourism medicine* is a multidisciplinary approach focused on protecting and improving health during travel, involving training, health promotion, and disease prevention through medical consultation.
- *Health tourism* involves travel to spas or destinations to improve physical conditions under medical supervision, providing physical exercise, therapy, and nutritional control (Skupnjak, 1996, 346).

The types of health tourism are as follows:

- *Medical tourism* involves traveling for healthcare (surgery, dental, cosmetic, psychiatric, and alternative treatments) in medical offices, clinics, or exceptional hospitals for rehabilitation because of various diseases (cardiovascular, respiratory, neurological, orthopedic, locomotor, and/or dermatological diseases).

- *Spa tourism* is conducted through the professional and controlled use of natural medicinal agents under professional supervision to preserve and improve health, which are often provided in rehabilitation hospitals or health hotels through various thalassotherapy treatments, pool facilities, thermal/sulfur/naphthalene/mineral waters, healing mud, and specially designed recovery programs.
- According to the World Tourism Organization and the European Travel Commission (2018), *wellness tourism* fosters individual well-being by promoting balance among the physical, mental, emotional, occupational, intellectual, and spiritual dimensions. It includes the organized implementation of health-preventive and curative programs with multiple disciplines (experts in physiotherapy, osteopathy, rehabilitation, kinesiology, nutrition, psychotherapy, and spa therapists). It implements complementary methods of conventional and alternative medicine (antistress and detox programs, life-coach consulting, beauty treatments, aromatherapy, saunas, massage, acupuncture, acupressure, body shaping, etc.), which are usually provided by health resorts.

Health tourism services are offered to domestic populations within the health and social insurance system, with partial reimbursement, and increasingly to foreign demand. Clients seeking health tourism services abroad are usually uninsured or seeking specific procedures not covered by their insurance.

Clients self-initiatively choose another country for a specific medical procedure in free circulation. In contrast, state medical tourism is linked to state or private health funds that refer their contributors to treatment or medical treatment in other countries because it saves money or is the only possibility for a specific procedure.

In Croatia, most wellness and a significant part of medical tourism are privately owned, market-oriented SMEs. In addition to hospitals, sanatoriums, and large hospital systems are often publicly owned and aimed mainly at users of the state health insurance institute (Ivandić et al., 2016, 32), limiting foreign tourist access due to capacity constraints. Through another prism, owing to more market-oriented tourism, health tourism services become available only to a paying clientele, denying necessary rehabilitation availability for individual crisis intervention conditions of health and social cases (postischemic conditions, bone and locomotor injuries, paresis, insults, dispossession). Combining market-oriented tourism with state health insurance systems can limit necessary rehabilitation services for local patients while creating a hospital-like atmosphere for tourists.

Although wellness and medical tourism may initially appear to occupy opposite ends of the tourism spectrum, both are experiencing growth and progress, garnering substantial global attention (Wright & Zascerinska, 2022).

Expansive Growth of the International Demand for Health Tourism Services

Health-related medical and wellness tourism are rapidly growing industries globally, with a 25% annual growth rate and a value exceeding 200 billion euros. Approximately 14 million people travel annually for medical reasons. Over 203 million trips are made annually at the European level, with approximately 115 billion Euro spent (American Chamber of Commerce in Croatia, 2018, 3). According to *the Medical Tourism Market—Global Industry Analysis Report*, the forecast by 2027 will be a value of 272.70 billion US dollars.

The growth in demand for medical tourism is driven by differences in health system organization and service prices, the avoidance of long queues, and the fact that standard health insurance often covers only essential services. Patients choosing medical tourism destinations focus on service standards, safety, quality, price, medical staff expertise, modern technologies, and destination attractiveness in cooperation with international agents/facilitators of medical tourism trips.

Therefore, travel agencies play crucial roles in promoting health, such as "International Medical Travel," "European Tours" or a specialized agency in the Republic of Croatia—"Croatia Medical Travel," offering "*all inclusive*" packages that include top medical facilities, internationally qualified experts, and organized travel and accommodations. Croatia, for example, promotes health tourism through a dedicated website, "*Croatia—full of well being,*" connecting private and public sector initiatives with potential foreign clients.

Medical tourism's impressive growth in income and the number of countries involved in joint entrepreneurial actions necessitate bioethical consideration and boundaries (Barač Miftarević, 2022). Since health tourism enters the border area of health and hotel-hospitality-tourism activities, it is expected that authorities will have to adopt special international regulations, standards, and accreditation algorithms to ensure patient safety and quality care.

Bioethical Aspects of Market Mechanisms of Health Tourism

The sudden expansion of health tourism has brought numerous benefits and disappointment to millions of patients in search of available medical procedures. Therefore, there is a need to regulate and supervise institutions to ensure and maintain the health and safety of health tourists.

The international trade in health tourism has opened up risk issues, unpredictable outcomes of individual procedures, the quality of health care, the informality of informed consent, the absence of regulatory frameworks, and the issue of cross-border responsibility.

Medical travel can have unintended, unwanted outcomes for patients seeking affordable healthcare. Lower costs sometimes mean lower-quality services (Turner, 2007). The greatest threat to using medical services in another country for the user is possible uncertainty, that is, the possibility of receiving inadequate health care compared with the expected one.

Some countries lack medical supervision, professional health personnel, and accreditation by international bodies such as JCI (Joint Commission International). The regulations that prescribe minimum standards, levels of arrangement, quality equipment, and verified programs are also inconsistent. Having only certificates such as established *ISO standards* that regulate business and hygiene procedures, some institutions lack market-relevant health and medical expertise certificates. Therefore, international quality standards modeled after the ISQUA (International Society for Quality in Health Care) and HCQI (Healthcare Quality Indicator) should be set, with the World Health Organization ensuring user safety at the international level but also ensuring the monitoring of services in national tourism.

The establishment of a destination management system at the level of tourist destinations that develops medical tourism based on the principles of IQM (*integrated quality management*) should represent a further step in establishing a standard development model of medical tourism for every country that develops this form of tourism.

A forthcoming European Commission Directive aims to define harmonized patient rights and equal health service opportunities in Europe, enhancing cooperation between EU countries. Croatia's Ministry of Health has an Institute for Health Services in Tourism, which specializes in performing a wide range of professional tasks related to the development of medical tourism and, in general, in increasing the quality of health services in tourism.

However, not all countries adhere to quality standards or unified biolegislative policies.

A particularly significant bioethical issue in medical tourism is related to certain medical services that are not allowed in the countries of residence. Specifically, the motive for using medical services abroad is often the use of new technologies, innovative health procedures, and even experimental drugs (Goldbach & West, 2010), as well as medical procedures that are not provided in the country of residence or are not allowed by legal regulations.

"*Therefore, travel to countries where medical procedures and treatments are cheap, the ability to skip queues or obtain affordable organs for transplantation, access to miraculous rejuvenation treatments,* etc., *are simply transferred from rigor to the realm of transactions*". Therefore, there are social, ethical, legal, and health consequences of unacceptable behavior, such as consuming health tourism without regulatory and ethical frameworks (Vuletić, 2022).

It is also assumed that procedures and treatments at affordable prices outside the country's borders and the unacceptable inclusion of medical care in the mercantile profit and economic efficiency framework will worsen the already endangered healthcare system. A reduction in health benefits offered by governments and employers will likely increase the number of individuals seeking affordable medical care in the global market for privatized, commercial health care (Turner, 2007).

In recipient countries, there is also a bioethical problem of health inequality in the distribution toward local patients, especially those who depend on public health care, since medical tourists use scarce resources that are then lacking in the local population.

Medical tourism is increasingly governed by market forces, transforming healthcare services into tradable assets and thereby reducing human attributes to goods that can be bought, sold, or bargained over. This shift poses a serious threat to modern bioethical principles, as market mechanisms often achieve efficient allocation but do not necessarily uphold ethical standards (Badulescu & Badulescu, 2014).

Furthermore, health and medical care cannot be treated as mere commodities tied to a price tag (Pellegrino, 1999). Such a perspective is ethically untenable and detrimental to patients, healthcare professionals, and society at large (Connell, 2006). Given that healthcare is both a universal human need and a collective good, it should be made accessible to all citizens through equitable distribution.

Conclusion

The promotion of medical services and health tourism is an important strategy for the economic development of many countries. However, this initiative should not be under the radar of medical standards or exceed legal and bioethical frameworks. To meet the quality standards in health tourism, it is essential to invest in hotel and healthcare infrastructure, which aligns with the demands of modern global tourism. This requires staff training and efforts to enhance competitiveness in the international market, ultimately ensuring a consistent and high-quality service offering. The realized economic income should be redirected to increasing the social fund for health services of the domicile population, reducing payment conditions for clinically necessary cases, and increasing the spatial capacity of spa tourism. This approach would enhance health care and uphold bioethical principles of equal availability and fair distribution, ensuring that fundamental human rights are met.

References

American Chamber of Commerce in Croatia (2018). *Preporuke za razvoj medicinskog turizma u Hrvatskoj*. Retrieved from https://www.amcham.hr/storage/upload/doc_library/preporuke_za_razvoj_medicinskog_turizma_u_hrvatskoj_104329.pdf

Badulescu, D., & Badulescu, A. (2014). Medical tourism: Between entrepreneurship opportunities and bioethics boundaries. *Iranian Journal of Public Health., 43*(4), 406–415.

Barač Miftarević, S. (2022). Medical tourism in Croatia: Where are we now? *Journal of Applied Health Sciences, 1*(8), 121–131. https://doi.org/10.24141/1/8/1/11

Carrera, P. M., & Bridges, J. F. (2006). Globalization and healthcare: Understanding health and medical tourism. *Expert Review of Pharmacoeconomics & Outcomes Research, 6*(4), 447–454. https://doi.org/10.1586/14737167.6.4.447

Connell, J. (2006). Medical tourism: Sea, sun, sand and surgery. *Tourism Management*, *27*(6), 1093–1100. https://doi.org/10.1016/j.tourman.2005.11.005

Goldbach, A. R., & West, D. J. (2010). Medical tourism: A New venue of healthcare. *Journal of Global Business Issues*, *4*(2), 43–49.

Han, H., & Hyun, S. S. (2015). Customer retention in the medical tourism industry: Impact of quality, satisfaction, trust, and price reasonableness. *Tourism Management, 46*, 20–29. https://doi.org/10.1016/j.tourman.2014.06.003

Ivandić, N., Kunst, I., Telišman–Košuta, N. (2016). *Pretpostavke održivosti zdravstvenog turizma u Republici Hrvatskoj*. Varaždin.

Menvielle, L., Menvielle, W., & Tournois, N. (2011). Medical tourism: The model of making decisions in choosing the services. *Turizam*, *59*(1), 45–59.

Müller, H., & Kaufmann, L. E. (2001). Wellness tourism: Market analysis of a special health tourism segment and implications for the hotel industry. *Journal of Vacation Marketing, 7*(1), 5–17. https://doi.org/10.1177/135676670100700101

Pellegrino, E. E. (1999). The commodification of medical and health care: The moral consequences of a paradigm shift from a professional to a market ethic. *Journal of Medicine and Philosophy, 24*(3), 243–266. https://doi.org/10.1076/jmep.24.3.243.2523

Skupnjak, B. (1996). Izglednost razvitka i konceptualno-organizacijski okvir zdravstvenog turizma u nas. *Tourism and Hospitality Management*, *2*(2), 343–356. https://doi.org/10.20867/thm.2.2.11

Špoljar, S. (1994). Prilog promidžbi zdravstvenog turizma Republike Hrvatske. *Acta Turistica, 2*, 253–264.

Turner, L. (2007). First world health care at third word prices: Globalization, bioethics and medical turism. *BioSocieties, 2*(3), 303–325. https://doi.org/10.1017/S1745855207005765

Vuletić, S. (2022). Moralno-bioetička problematika eksploatacijskih tendencija trgovanja ljudima. In S. Vuletić & V. Dugalić (Eds.), *Prevencija trgovanja ljudima* (pp. 39–68). Đakovo.

World Tourism Organization and Europea Travel Commission. (2018). *Exploring Health Tourism. Madrid: UNWTO*. https://doi.org/10.18111/9789284420209

Wright, D. W. M., & Zascerinska, S. (2022). Becoming immortal: Future wellness and medical tourism markets. *Journal of Tourism Futures*. https://doi.org/10.1108/JTF-05-2021-0119

CHAPTER 19

Conclusion

Ivan Pavić, Nikša Alfirević, and Suzana Vuletić

Abstract As technology and innovation accelerate, challenges that were once local have rapidly scaled into pressing global concerns, affecting social equity, environmental health, and the development of humanity. It has become clear that confronting these issues requires more than a single (technological, economic, or a social) viewpoint. Bioethics offers an interdisciplinary approach, uniting philosophical and moral analysis to examine the ethical implications inherent in ongoing technological and industrial advancements.

Keywords Bioethics · Biotechnology · Sustainability · Equity

I. Pavić
University of Split, Split, Croatia
e-mail: ivan.pavic@efst.hr

N. Alfirević
Faculty of Economics, Business and Tourism, University of Split, Split, Croatia
e-mail: nalf@efst.hr

S. Vuletić (✉)
Catholic Faculty of Theology in Đakovo, Josip Juraj Strossmayer University of Osijek, Osijek, Croatia
e-mail: suzanavuletic007@gmail.com

I. Pavić et al. (eds.), *Bioethics and Social Ethics in The Modern World*,
https://doi.org/10.1007/978-3-031-86418-6_19

Developing biotechnology and other scientific innovations raises questions about social implications, environmental sustainability, and broader implications for humanity. Issues that once appeared to be local or isolated have increasingly become global challenges, requiring scientific and technical solutions and thoughtful ethical and philosophical considerations.

Addressing the issues of social equality and environmental challenges becomes increasingly different by using a single perspective—technical, economic, or social. There is a need to integrate multiple views and consider different implications of the current science and technology. Bioethics represents an excellent theoretical framework for such a task, as it offers an integrative framework, incorporating philosophical and moral reflections of the contemporary world.

Framing the questions of sustainable development in the twenty-first century using the theoretical lens of bioethics could help evaluate the ethical implications of technological and industrial progress and set transparent limits that balance innovation with protecting environmental and social resources. This could lead to developing new theoretical models and actionable agendas, ensuring equitable resource distribution, stable economic development, and the protection of the environment, and helping us balance the interests of present and future generations.

Ultimately, a sustainable society is based on ethics and appreciation of life in all its forms and diversity. We hope this book contributes to promoting such an understanding among its readers.

Index

A

abortion, 64, 72–74, 76–82, 84–93, 211
agenda setting, 118, 119, 121, 123, 124
Aquinas, Thomas, 38, 40–44
Aristotle, 38–41, 43, 52

B

bioethical principles, 7, 9, 26, 66, 211, 215, 216
bioethics, 2, 3, 6, 8–12, 16, 24–27, 29, 30, 32, 33, 43, 178, 179, 220

C

Catholic Church, 74, 79, 182, 206
Church, 74, 180, 181, 205, 206
Complementary and alternative medicine (CAM), 106, 107, 109–113
Constitutional Court of Croatia, 72, 73, 80–82, 86, 87, 90
COVID-19, 105, 107, 109–111, 113, 188, 192

E

ecology, 179–181, 183, 184
economic crises, 188, 191–196
environment, 6–8, 12, 116, 117, 122, 123, 136, 143, 149, 150, 153, 155, 160, 180, 182, 188–192, 194, 202, 206, 207, 220
environmental impact, 141, 155, 156, 159, 195, 204
environmental protection, 6, 188, 196, 205, 206
ethical implications, 2, 106, 179, 220
ethical regulation, 26
ethics, 2, 3, 6, 9–11, 16, 28, 32, 33, 48, 49, 52, 53, 56, 57, 116, 117, 135–137, 139–141, 143, 179, 180, 189, 196, 220
European Court of Human Rights (ECtHR), 63, 81, 101
evil, 16–21, 54

I. Pavić et al. (eds.), *Bioethics and Social Ethics in The Modern World*, https://doi.org/10.1007/978-3-031-86418-6

F
food consumption, 148–153, 155–160

G
Gender Identity Disorder (GID), 98, 100
goal of medicine, 50, 53
God, 19, 41, 42, 44, 180–182, 184, 205–207

H
health protection, 90
health tourism, 210–216
human dignity, 11, 19, 30, 67, 93, 204
Human Evilization, 19
human rights, 30, 32, 33, 62, 67, 75, 78, 81, 83, 88, 91, 93, 190, 196, 216

I
individualism, 136, 203
integral bioethics, 38, 43, 44
integrative model, 6

L
Laudato Si, 183, 184, 201, 202, 206
legal regulation, 24, 26, 28, 61, 211, 215
LOHAS consumers, 134, 135, 138

M
mass production, 18
materialism, 136, 203
medical ethics, 32, 48, 50, 51, 53, 54, 82–84
medical tourism, 100, 210–215
megatrends, 134, 143
motherhood, 63, 67

N
natura integra, 38, 43, 44

P
parenting, 75, 136
patient's good, 50–52, 54
Pellegrino, Edmund, 48–57, 216
philosophical ethics, 48
Pope Francis, 181, 183, 184, 201–204

Q
quality of medical service, 210

R
reproductive rights, 81
responsible marketing, 134, 135, 139

S
Sex Reassignment Surgery (SRS), 98–101
social ethics, 2, 6, 10–12, 189
social media, 122
St. Francis, 182–184, 207
surrogacy, 60–64, 66, 67, 211
sustainability, 2, 3, 6–12, 33, 119, 120, 134, 138, 139, 141, 149–153, 160, 172, 181, 188–191, 194–196, 220
sustainable behavior, 2, 152, 158
sustainable development, 8, 12, 188, 189, 196, 220
Sustainable Development Goals (SDGs), 7, 139, 149

T
technological imperative, 200
theology, 83, 84, 180, 181
tradition, 29–31, 52, 157, 168, 173

V
virtue ethics, 7, 10, 48

W
well-being, 7, 10, 53, 151, 158, 159, 190, 196, 210–212
World Youth Day (WYD), 201, 202, 204–206

Kolibri 360 GmbH
Ferdinand-Jühlke-Straße 7
99095 Erfurt, Deutschland
produktsicherheit@kolibri360.de